Safari Guides

Larger Animals of East Africa

David Hosking & Martin B. Withers

HarperCollins*Publishers*

FOR JEAN, SIMON & MARK

HarperCollins*Publishers*
77–85 Fulham Palace Road
London
W6 8JB

The HarperCollins website address is:
www.**fire**and**water**.com

Collins is a registered trademark of HarperCollins*Publishers* Ltd

10 9 8 7 6 5 4 3 2

03 02 01 00 99

ISBN 0 00 220036 8

Colour reproduction by Colourscan, Singapore

Printed and bound by Rotolito Lombarda, Italy

ACKNOWLEDGEMENTS

There are countless people to whom the authors owe a debt of gratitude for the production of this book. They would like to express special thanks to Myles Archibald and his staff at HarperCollins. To the staff at United Touring International Ltd, London, in particular Joanne Peers, Versha Veghala, Cordelia Robb, Helen Simmans and Jonathan Bolden. To Neil Outram, General Manager United Touring Company, Arusha, Tanzania and to John Glen, General Manager United Touring Company, Nairobi, Kenya. We would like to make special mention of our drivers Joseph Nicholas Kwelukilwa, Moses Sengenge and Suleiman Kiruwa in Tanzania and Dominic Gichinga in Kenya without whose driving skills and sharp eyesight many of the illustrations featured would not have been obtained. Thanks are also due to the management and staff of the Frank Lane Picture Agency, Graham Armitage CZ Scientific Instruments Ltd, Graham Rudaford Fuji, London, Phil Ward, Dr D.A.P.Cooke, Dr. Gordon Reid and his staff at Chester Zoo and all clients of Hosking Tours who have helped to make much of our time in East Africa such an enormous pleasure.

Lastly, we owe a huge thank you to our wives and families for their support and tolerance during the production of this book.

Martin B. Withers FRPS
David Hosking FRPS

PREFACE

The main objective of this book is to provide visitors to East Africa with a guide to the larger mammals most likely to be encountered during the course of a safari in and around the main National Parks and Game Reserves of Kenya, Tanzania and Uganda, although many of the species featured are also commonly found in adjacent countries. It was not our intention to produce complete coverage of the mammals of East Africa. The species included in this edition were selected based on the personal experiences of the authors, gained during visits to the region over a period of almost twenty years. Anyone wishing to obtain greater and more detailed information on species should consult the bibliography.

An attempt has been made to provide as full a diagnostic description of each species featured as space would permit and to highlight key facts about each species for quick reference. The illustrations have been selected to show the main features of identification.

BEFORE YOUR DEPARTURE

As a general rule it is unwise to commence a photographic safari with any untested equipment. Always put at least one roll of film through a new camera and carry out a full test on any newly purchased or untried lens. You should also endeavour to familiarise yourself with the use and functions of all equipment, accessories and film stock prior to your departure.

Time spent researching the areas you will be visiting can prove invaluable and, having established the flora and fauna you are likely to encounter, will aid you in selecting the most appropriate equipment to satisfy your own photographic needs.

It is also good practice to check that you have adequate insurance cover for all your equipment. Ensure that your policy fully covers equipment for loss or damage in all the countries you plan to visit and for the duration of your tour.

CARE & MAINTENANCE

As a general rule it is advisable to thoroughly check and clean all your camera bodies and lenses at the end of each day's shooting. All equipment used on safari is subject to the potentially damaging effects of sunlight, damp, rain and dust. Do remember to keep camera bags and film stock out of direct sunlight whenever possible. Dust is probably the most damaging of all, a single grain on the camera pressure plate or, in the jaws of a film cassette, can badly scratch a complete roll of film. A rubber blower brush is ideal for keeping the inside of your cameras clean, while lens elements are best cleaned with specially purchased cleaning fluid and tissues. The outer casings of both cameras and lenses can best be cleaned using an ordinary household paint brush. It is always worth having a supply of large plastic bags with you, into which you can seal your entire camera bag on the days when you are travelling from one location to another. This will greatly reduce the risk of dust entering the most sensitive parts of your cameras.

GETTING CLOSE TO BIRDS AND MAMMALS

Whilst on safari the vast majority of your photography will be undertaken from a vehicle during game drives. Many opportunities also exist for wildlife photography on foot, within the grounds of safari lodges and at specially designated areas within the National Parks and Reserves. At many of the lodges getting close to birds is often quite easy, due to the tame nature of many species. Others, however, require some knowledge of basic 'stalking' procedures to gain a close enough approach for a worthwhile image size in the finished pictures. Avoid walking straight towards your intended subject, as this is likely to cause it to fly away, a slow angled approach is more likely to succeed. Watch your subject during your approach and should it appear concerned 'freeze' for a while until it again looks

relaxed. A crouched approach during a 'stalk' can also be beneficial, the smaller you appear the less likely you are to frighten your intended subject. It is also a good idea to make use of any natural cover that the terrain may offer.

Many of the safari lodges in East Africa provide food and water for the local bird population and these feeding areas can offer good photographic opportunities. Even if the bird tables themselves are non-photogenic you can attempt to photograph birds on natural perches as they make their way to and from the feeding areas.

You should exercise considerable caution when 'stalking' mammal species, remember they are wild animals and can be extremely dangerous. Never try to tempt monkeys or baboons to come closer with food items - this is a sure way to get badly bitten.

CODE OF CONDUCT

The National Parks and Reserves of East Africa operate a strict Code of Conduct for both drivers and visitors. A short, simple list of do's and don'ts have been implemented to minimise disturbance to the birds and mammals, to lessen the impact of tourism on the environment and to ensure that all visitors experience safe and enjoyable safaris.

Please do not pressurise your driver into breaking Park regulations, you will be jeopardising his job and run the risk of expulsion from the Park.

Please keep noise to a minimum, particularly when close to animals and never leave the vehicle, sit or stand on the roof, or hang precariously from the windows.

Never discard any form of litter, apart from being unsightly it can cause serious injury or even kill animals if ingested.

Cigarettes are best avoided during game drives, the careless or accidental discarding of a match or cigarette stub can lead to uncontrollable fires, resulting in the deaths of many living creatures.

GENERAL NOTES

Many people visiting East Africa express a wish to photograph the local people. Before doing so please obtain permission and be prepared for the possibility of paying for the privilege. On no account attempt to photograph military installations or personnel.

PICTURE CREDITS

The authors would like to thank the Frank Lane Picture Agency for their assistance in the compilation of the photographs used in this book. All the photographs are by the authors with the exception of those listed below:

Giant Forest Hog	R.Austin
Common Duiker - adult	P.Perry
Sitatunga - male	J.Tinning
Lesser Kudu	L.Batten
Bongo	R.Austing
Bongo	F.Lane
Tree Hyrax	M.Newman
Side-striped Jackal	F.Lane
African Civet	F.Lane
Patas Monkey	F.Lane
Aardwolf	F.Lane
Aardwolf	W.T.Miller

BIBLIOGRAPHY

The following books are recommended for the purposes of further reference:

Apps, Peter, *Wild Ways,* Southern Book Publishers 1992.

Dorst, J. & Dandelot, P. *A Fieldguide to the Larger Mammals of Africa.* HarperCollins.

Eley, R.M. *Know your Monkeys - A Guide to the Primates of Kenya.* National Museums of Kenya.

Estes, R.D. *The Safari Companion - A Guide to Watching African Mammals.* Tutorial Press.

Haltenorth, T. & Diller, H. *A Fieldguide to the Mammals of Africa including Madagascar.* HarperCollins.

Kingdon, J. *East African Mammals.* 7 vols. The University of Chicago Press.

Kruuk, H. *The Spotted Hyena.* The University of Chicago Press.

Norton, B. *The Mountain Gorilla.* Swan Hill Press.

Simon, N. *Between the Sunlight & the Thunder.* HarperCollins.

Spinage, C.A. *The Natural History of Antelopes.* Croom Helm.

Bush Pig

Order	Family	Genus & Species
Artiodactyla	*Suidae*	*Potamochoerus porcus*

Identification - A typical hog, with a long face and a short rotund body. The coat is long and coarse and varies in colour from reddish to dark brown. Along the back the hair is longer, forming a dorsal crest. The head has a conspicuous pattern of black and white markings which are extremely variable and in some instances the whole head appears white. The tail is long and thin, measuring up to 40 cms. in length. The hooves have four toes. Very young animals have a spotted pelage of buff on dark brown, but these spots are replaced, at about six months old, by a rufous brown coat and the longer hair of the dorsal crest. Males are larger and heavier than females.

Distribution & Habits - An animal of forests, woodlands, thick bush, and other areas offering sufficient cover and water. Bush Pigs are gregarious animals living in family groups usually consisting of a boar with several sows and offspring, numbering on average 15 to 20 pigs in all, although congregations of over 40 animals have been recorded. The boar is the dominant family member, leading and protecting his group. The sows give birth at anytime of the year in a nest prepared in advance, usually in dense vegetation. The sow will keep the piglets well hidden for the first two months of their lives, leaving them in the nest while away foraging. Bush Pigs are omnivorous, eating a wide variety of vegetation as well as insects, reptiles, birds eggs, seeds and fruits which they search out by rooting around and digging with their snouts. Occasionally their feeding habits bring them into conflict with man, as a foraging group can do vast damage to crops.

KEY FACTS

Size:	Height: 54 - 80 cm.
	Weight: 70 kgs (average).
Breeding:	Gestation: About 4 months.
	Young: Litters of 3 - 6.
	Sexual Maturity: Male 24 months, female 21 months.
	Births: No specific season.
Lifespan:	12+ years.
Lifestyle:	Family: Average size 15 - 20.
	Diet: Omnivorous.
	Main Predators: Leopard, Spotted Hyaena & Lion.
	Habitat: Varied. Forests, woodlands, bush and scrub.
Conservation & Status: Common, not endangered.	

Giant Forest Hog

Order	Family	Genus & Species
Artiodactyla	*Suidae*	*Hylochoerus meinertzhageni*

Identification - The largest of East Africa's pig species. A huge, heavily built animal with long black body hair. The head is large, with patches of light hair on the long face. They have a broad snout and large distended ridges below the eyes. Mature males are much larger than females and have tusks which can be as long as 30 cms, emerging in a backward direction from the mouth, the tusks are much smaller in the female. The ears are small and pointed. The young are uniform brown in colour. The tail is long and thin, measuring up to 45 cms. in length.

Distribution and Habits - An animal usually associated with dense forest and very often those of high mountains. Mainly nocturnal, they usually remain in deep cover during the daytime and are most likely to be seen foraging at dusk in forest glades. They feed on grasses, sedges, leaves, berries and fallen fruits. Although they will eat certain plant roots they do very little digging or rooting for food with the snout. They live in family parties of 4 to 12 animals, consisting of a mature boar, a sow and several generations of offspring. Sows give birth to average litters of 2 to 6 piglets in a nest of dry grass and other vegetation prepared in advance and sited in dense cover. At birth the young are uniform buff in colour, however, the fur soon darkens to brown, becoming black on reaching full maturity.

KEY FACTS

Size:	Height: 85 - 100 cm.
	Weight: Males 230 kgs. (average).
	Females 180 kgs. (average).
Breeding:	Gestation: 4 months.
	Young: Average litters 2 - 6.
	Sexual Maturity: Males 3 - 4 years.
	Females 1 year.
	Births: No specific season.
Lifespan:	12+ years.
Lifestyle:	Family: Small parties of 4 - 12.
	Diet: Herbivorous.
	Main Predators: Leopard, Spotted Hyaena.
	Habitat: Forests.

Conservation & Status: Numbers are difficult to determine but they are not thought to be endangered. Numbers can, however, fluctuate greatly as they are susceptible to swine fever and other diseases.

Warthog

Order	Family	Genus & Species
Artiodactyla	*Suidae*	*Phacochoerus aethiopicus*

Identification - Often regarded as ugly, the Warthog is the most common of the African pigs. They have little in the way of fur, just a few bristles and whiskers on the body of grey skin. They do, however, have a long black mane of hair on the neck and shoulders. Coloration can vary greatly due to their habit of wallowing in muddy pools. The tail is long and thin, measuring up to 50 cms. in length and is carried vertically when running. They have a large flat face on which are found two sets of 'warts', one set immediately below the eyes and the other on the sides of the face between the eyes and the mouth. They have tusks, which emerge from the mouth in a semi-circle outwards and upwards. the tusks and warts are less prominent in the sow than in the boar.

Distribution & Habits - An animal of open savannahs and woodlands, living in family groups consisting of a boar, a sow and the offspring from several litters. They can be found during daylight hours grazing, which they often do while kneeling on their front legs. During the hottest part of the day they will seek shade or cover. They feed mainly on short grasses, but will take leaves, roots, fruits and tubers. Sows give birth in burrows and hollows which are lined with grass. The young will remain in the nest for 6 - 7 weeks. Whole families will sleep together in burrows often excavated originally by Aardvark and later enlarged by the hogs. To defend themselves against predators they enter the burrow backwards, this enables them to make good use of their formidable tusks if threatened.

KEY FACTS

Size: Height: 65 - 85 cm.
        Weight: Males 85 kgs (average).
        Females 57 kgs (average).

Breeding: Gestation: 5.5 months.
        Young: The average litter is 3 or 4.
        Sexual Maturity: Boars 2 years,
        Sows 18 months.
        Births: Peaks in March/April.

Lifespan: Up to 18 years.

Lifestyle: Family: Lives in family parties.
        Diet: Vegetation.
        Main Predators: Lion. Leopard and Cheetah.
        Habitat: Mainly open savannah & woodland

Conservation & Status: Very common.

Hippopotamus

Order	Family	Genus & Species
Artiodactyla	*Hippopotamidae*	*Hippopotamus amphibius.*

Identification - Second in weight only to the Elephant, this unmistakeable amphibious mammal has a bulbous body, short legs and a large head which broadens at the muzzle. The eyes, ears and nostrils are placed high on the head in order to remain clear of the water when the animal has it's body submerged. The coloration is pink, grey/purple and brown. The body is devoid of fur, having just a few bristles on the tail, head and face. The tail is short and thick. They have well developed incisor teeth which are used when fighting and serve no purpose at all with regard to feeding, which is accomplished by use of the large lips in a ripping motion.

Distribution & Habits - An animal of rivers and swamps across much of East Africa. During the daytime they usually remain partially submerged to avoid the effects of overheating, sunburn and dehydration. At night time they leave the water to graze, preferring short grassy pastures. They have been recorded, in isolated cases, feeding on the rotting flesh of other animals. They use well defined pathways from the water to their feeding areas, and during the course of the night's foraging, they may cover a distance of 5 to 6 miles. They will usually gather together in herds of 10 to 50 animals, but during periods of drought densities can increase dramatically to 200 or more. Adults are capable of remaining completely submerged for periods of up to 5 minutes. When giving birth cows will isolate themselves from the herd, remaining alone with their calves for about 6 weeks. Calves are capable of suckling underwater as well as on land.

KEY FACTS

Size:	Height: 130 - 165 cm.
	Weight: Males 1480 kgs (average).
	Females 1360 kgs (average).
Breeding:	Gestation: 8 months.
	Young: One only.
	Sexual Maturity: Male 7 - 12 years,
	Female 7 - 15 years.
	Births: Usually in the dry seasons,
	calves produced at 2 year intervals.
Lifespan:	Around 30 years.
Lifestyle:	Family: Groups of 10 - 50 on average.
	Diet: Herbivorous.
	Main Predators: Young calves are taken
	by Lions and Crocodiles.

Conservation & Status: Still relatively common throughout East Africa.

Masai Giraffe

Order	Family	Genus & Species
Artiodactyla	*Giraffidae*	*Giraffa camelopardalis*

Identification - As the world's tallest animal the giraffe is unmistakeable. The immense neck, sloping body and long legs aid easy identification. The coat pattern is that of irregular brown blotches on a yellow-buff background. The coat colour of the males has a tendency to darken with age. The underparts are light with faint blotches and spots. A mane of stiff hair extends from the nape, down the neck to the shoulders. The tail is long and thin, terminating in an abundance of long black hair. The amount and size of horns is very variable, but normally they possess a principle pair on the upper forehead and signs of a much smaller pair on the crown. In addition they often have a single knob of horn in the centre of the lower forehead. The horns are skin covered with tufts of hair at the tips. Females are smaller than males.

Distribution & Habits - An inhabitant of bush and lightly wooded regions south of the Sahara, where they feed by browsing to a height denied to all other herbivores. They will strip leaves and shoots from even the most thorny trees and bushes with ease, by use of prehensile lips and a very long 45 cm. tongue. They remain active for much of the day, seeking shade during the hot midday period. They are found in herds of between 6 - 12 animals on average. They need to drink every 2 - 3 days when water is available and during the dry season they are seldom too far from a permanent source, they will disperse over a much wider area during the rains. In order to drink they have to splay out the forelegs and it is at these times they are most susceptible to attack by lions. They walk with a slow amble moving both legs of the same side of the body together, this is known as 'pacing' and is quite unusual among animals. When they run or gallop they give the impression of 'slow motion' but are capable of speeds up to 60 kph.

KEY FACTS

Size	Height: 4.25 - 5.5 metres.
	Weight: Males 1100 kgs (average),
	Females 700 kgs (average).
Breeding	Gestation: Around 14 months.
	Young: One usually, twins rare.
	Sexual Maturity: Males 3 - 3.5 years,
	Females 2.5 - 3 years.
	Births: No specific season.
Lifespan:	25+ years.
Lifestyle	Family: Small herds of 6 - 12 on average
	Diet: Herbivorous.
	Main Predators: Lions, but rarely taking
	full adults, and Spotted Hyaenas who prey
	on young.
	Habitat: Bush & lightly wooded regions

Conservation & Status: Numbers are stable, following a decrease over many years due to poaching for the hairs of the tail.

Reticulated Giraffe
A sub-species of Masai Giraffe

Order	Family	Genus & Species
Artiodactyla	*Giraffidae*	*Giraffa camelopardalis reticulata.*

Identification - The tallest of the world's animals, standing over 5 metres makes identification easy. They have an extremely long neck, sloping body and long legs. The coat pattern is very striking. Crisp, liver-coloured geometric patches, conspicuously defined by narrow white lines running between them, produces a 'crazy paving' appearance. These markings become paler on the inner flanks, legs and underparts. The tail is long and thin terminating in a profusion of long black hair for which it is often poached. Horn size and number varies greatly but usually a principle pair are visible on the upper forehead with a smaller set on the crown. In addition they often have a single knob of horn in the centre of the lower forehead. A mane of stiff hair extends down the back of the neck from the nape to the shoulders. The coat colour of the male has a tendency to darken with age. Females are smaller than males. A distribution overlap brings Reticulated Giraffes into contact with Masia Giraffes resulting in a variety of hybrid types within certain areas.

Distribution & Habits - An inhabitant of dry bush country in north-eastern Kenya and Somalia, feeding by browsing to a height denied to all other herbivores. They strip leaves and shoots from even the most thorny of trees and bushes with ease, by use of prehensile lips and a very long 45 cms. tongue. They usually remain close to a source of water and, if possible, will drink every 2 to 3 days. They can, however, go without water for considerable periods, obtaining sufficient fluid from their food intake. They remain active for much of the day, seeking shade during the hot midday period. They are usually found in herds of 6 to 12 individuals. They walk with a slow amble moving both legs on the same side of the body together, this is known as 'pacing' and is quite unusual among animals. When they run they appear to move in 'slow motion' but are capable of speeds up to 60 kph.

KEY FACTS

Size	Height: 4.25 - 5 metres.
	Weight: Males 1100 kgs (average), Females 700 kgs (average).
Breeding	Gestation: Around 14 months.
	Young: One only.
	Sexual Maturity: Males 3 - 3.5 years,
	Females 2.5 - 3 years.
	Births: No specific season.
Lifespan:	25+ years.
Lifestyle	Family: Small herds from 6 - 12 animals.
	Diet: Herbivorous.
	Main Predators: Lion, but rarely taking adults, and Spotted Hyaena who prey on young.
	Habitat: Dry bush country.

Conservation & Status: May be decreasing as a result of poaching.

Rothschild's Giraffe
A sub-species of the Masai Giraffe

Order	Family	Genus & Species
Artiodactyla	*Giraffidae*	*Giraffa camelopardalis rothschildi.*

Identification - The extremely long neck, sloping body and long legs make this, the tallest animal in the world, one of the easiest to identify. A thicker set giraffe than the Masai Giraffe, of which it is a sub-species. The coat pattern is similar to the Reticulated Giraffe but not quite so well defined. There is no patterning at all below the knees or hocks, these areas being almost pure white. The body patterning, although fainter, extends to the inner flanks, upper legs and underparts. A mane of stiff hair extends from the nape to the shoulders. The tail is long and thin terminating in a profusion of long black hair. The number and size of horns is variable, but normally there is a principle pair on the forehead with another smaller set on the crown. In addition there is often a single knob of horn in the centre of the lower forehead. Females are smaller than males.

Distribution & Habits - An animal of bush and lightly wooded areas of western Kenya, where it feeds by browsing to a height denied to other herbivores. They are able to strip leaves and shoots even from the most thorny of trees and bushes, by use of prehensile lips and a very long 45 cm. tongue. They remain active throughout the day, usually seeking shade during the hot midday period. They will drink every 2 to 3 days when water is available, but are able to survive without water for considerable periods, obtaining sufficient liquid from their food intake. They walk in a slow amble moving both legs on the same side of the body together, this is known as 'pacing' and is quite unusual among animals. When they run or gallop they give the appearance of moving in 'slow motion' but are capable of speeds up to 60 kph.

KEY FACTS

Size	Height: 4.25 - 5.5 metres.
	Weight: Males 1100 kgs (average),
	Females 700 kgs (average).
Breeding	Gestation: Around 14 months.
	Young: One only.
	Sexual Maturity: Males 3 - 3.5 years,
	Females 2.5 - 3 years.
	Births: No specific season.
Lifespan:	25+ years.
Lifestyle	Family: Herds of 6 -12 on average.
	Diet: Herbivorous.
	Main Predators: Lion, but rarely taking
	adults, Spotted Hyaena prey on young.
	Habitat: Bush and lightly wooded regions.

Conservation & Status: There has been a decrease in numbers during recent decades due to poaching, often for the tail hair.

Common Duiker

Order	Family	Genus & Species
Artiodactyla	*Bovidae*	*Sylvicapra grimmia.*

Identification - A medium-size, slender Duiker, with long legs, large ears and pointed horns. The horns are usually only present in the male, ranging from 7 to 18 cms in length. The pelage coloration ranges from grizzled grey to yellowish brown with a black streak running from the top of the russet forehead to the nose. The underparts are white tinged with grey and the legs are grey having a black band just above the hooves. The short tail is black above and white below. Females are generally larger than males.

Distribution & Habits - Mainly found in woodland habitats with scattered bush and scrub, avoids arid desert regions, as well as areas of open plains and dense forest. They feed mainly on the leaves and shoots of bushes, as well as tree bark, fruits and seed pods, rarely do they eat grass. They also consume insects and other animal matter from time to time, including frogs, small mammals and birds. They obtain the vast majority of their water requirement from their food and are able to survive for very long periods without drinking. Females give birth to single calves in dense cover, the newborn are able to run within 24 hour. They remain in dense cover sucking from the female 2 or 3 times a day growing very rapidly and by 6 months are as large as adults.

KEY FACTS

Size:	Height: 45 - 70 cm.
	Weight: Male 18.5 kgs (average).
	Females 21 kgs (average).
Breeding:	Gestation: 5 - 6 months.
	Young: One only.
	Sexual Maturity: Males and females from 1 year.
	Births: No specific period, peak during
	the rains. Females sometimes calve twice in a year.
Lifespan:	Up to 12 years.
Lifestyle:	Family: Pairs.
	Diet: Omnivorous.
	Main Predators: All mammal predators.
	Large eagles.
	Habitat: Woodland, bush and scrub.

Conservation & Status: Quite common in favoured areas, although they are hunted both for their flesh and by farmers suffering crop damage.

Kirk's Dik Dik

Order	Family	Genus & Species
Artiodactyla	*Bovidae*	*Madoqua kirkii*

Identification - The commonest Dik Dik in East Africa. A small delicate antelope, with a grizzled grey and brown coat, the legs are rufous/grey. The head, neck and shoulders sometimes show a flush of rufous brown. White patches encircle the large dark eyes and the males have short, spiky horns, measuring 6 - 11 cms in length. The elongated nose is a distinctive feature, serving as a very effective cooling device. They have prominent preorbital scent glands. Female Dik Dik are slightly larger than the males.

Distribution & Habits - A relatively common inhabitant of arid regions with bush or scrub cover. Dik Dik live in pairs and are very territorial, each male guarding his own 'patch' against other males. Territorial boundaries are marked by a succession of dung/urine sites, which are re-stated daily by both male and female. As well as dung sites the animals mark their territory by using preorbital scent glands. These glands exude a secretion which the animals spread on twigs, branches and grass stems. They are active during both the day and night, but seek shade during the hottest period of the day. They have extremely sensitive eyesight and hearing, seeking cover at the slightest sign of danger.

KEY FACTS

Size:	Height: 35-43 cms.
	Weight: Males 5 kgs (average).
	Females 5.5 kgs (average).
Breeding:	Gestation: 6 months
	Young: One only
	Sexual Maturity: Males 8-9 months, females 6-8 months.
	Births: Peaks in May and November.
Lifespan	Up to 9 years.
Lifestyle:	Family: Singles or in pairs.
	Diet: Herbivorous.
	Main Predators: Many, including Leopard, Cheetah, Serval, Jackal, Baboon and Eagles.
	Habitat: Bush and scrub.

Conservation & Status: Relatively common, but in many areas they face competition from domestic stock for food.

Steinbok

Order	Family	Genus & Species
Artiodactyla	*Bovidae*	*Raphicerus campestris*

Identification - A small, slender antelope with a reddish coat. The underside and the insides of the large ears are white. The horns, which are present only in the male, are straight spikes, measuring from 9 to 19 cms in length. They have long legs and the appearance of being higher at the hindquarters than at the shoulder. They have a pale patch around the eyes, contrasting with dark areas surrounding the preorbital scent glands. A triangular patch of black fur extends up the muzzle from the black nose. The young have a heavier, fluffier coat than the adults, but retain the same coloration.

Distribution & Habits - An animal of dry savannah and wooded hill sides, they graze on grass and browse trees, bushes and shrubs, as well as scraping with their hooves to expose roots and tubers. They are able to subsist without water, obtaining all their fluid requirement from their food. They are active during both the day and night in undisturbed areas, but will avoid the midday heat by seeking shade in regular resting places. Females will produce young throughout the year, the newborn, weighing about 1 kilo at birth, will suckle from it's mother almost immediately. The young commence grazing at about 2 weeks old and are weaned at 3 months. There are reports that from time to time Steinbok may take refuge underground in old Aardvark burrows, both for resting and for nursing young.

KEY FACTS

Size:	Height: 45 - 60 cms.
	Weight: 10.6 kgs (average).
Breeding:	Gestation: About 6 months.
	Young: One only.
	Sexual Maturity: Males 2 years, females 1 year.
	Births: No specific season
Lifespan:	10+ years
Lifestyle:	Family: Solitary or in pairs.
	Diet: Herbivorous.
	Main Predators: Vulnerable to all major predators.
	Habitat: Wooded hillsides & savannah.

Conservation & Status: Numbers are on the decrease as a result of suitable habitat being cleared by man for cultivation, domestic stock, roads, & buildings etc.

Oribi

Order	Family	Genus & Species
Artiodactyla	*Bovidae*	*Ourebia ourebi*

Identification - A small, slender built antelope with long legs and neck, the hindquarters are rounded and slightly higher than the forequarters. They have large pointed ears below which is situated a scent gland which appears as a patch of bare black skin. The coat is of silky fine hair, reddish to fawn above and white below. The chin is white and the dark eyes contrast strongly with the white eyebrows. The legs are reddish-fawn with tufts of slightly longer hair at the knee joints. The horns, which feature in the male only, have rings on the lower third, are curved backwards and grow to a length of 8-19 cms. The rufous tail is very short and has a black tip. Females are slightly larger than males.

Distribution & Habits - An antelope of open grassland regions, where they require short grasses for feeding, and taller grasses for cover while resting. The males will establish and defend territories forcefully, marking their areas by the use of six different scent glands found about the body. A female will hide her newborn in thick grass for the first 3 or 4 days of life, from then on, over a period of about a month, the calf will begin to spend more and more time accompanying it's mother. The young grow rapidly and attain full adult height at around 3 months. When alarmed the Oribi will emit a loud shrill whistle and will commence a 'stotting' performance, springing into the air with legs held rigid. Occasionally they will inflict considerable damage to cultivated crops, particularly during their night time feeding.

KEY FACTS

Size:	Height: 54 - 66 cms.
	Weight: 14 - 21 kgs.
Breeding:	Gestation: 7 months.
	Young: One only.
	Sexual Maturity: Males from 14 months,
	Females from 10 months.
	Births: Throughout the year, with peaks
	during the rains.
Lifespan:	About 14 years.
Lifestyle:	Family: Singly, pairs or family groups.
	Diet: Herbivorous.
	Main Predators: All mammal predators. Eagles
	and snakes will take young.
	Habitat: Prefers open grasslands.

Conservation & Status: The numbers of Oribi have dropped in recent years as a result of a loss of habitat brought about by new human settlements and agricultural development.

Klipspringer

Order	Family	Genus & Species
Artiodactyla	*Bovidae*	*Oreotragus oreotragus*

Identification - A small, strongly built antelope, with a wedge shaped head and a thick coat of olive-yellow and grey hair. The muzzle is washed with brown and the chin and underparts are white. The large rounded ears are white-lined with prominent black markings. They have dark preorbital scent glands which are very pronounced. The legs are sturdy, terminating in black hooves which are specially adapted for the rocky ground, giving the animal the appearance of 'standing on tiptoe'. The tail is very short and the horns are upright, measuring from 6 - 15 cms in length, occurring in the females of some populations as well as the males.

Distribution & Habits - An animal of steep, rocky hill sides and screes as well as isolated kopjes and high mountains, having been recorded as high as 4000m on Mount Kilimanjaro. They live in pairs or in small family parties and can often be seen standing atop a rock or boulder keeping a sharp lookout for predators. They are extremely agile, being capable of racing over boulder strewn terrain at high speed and with apparent ease. They feed on leaves, shoots and berries, taking fruits when available and very occasionally grass. The young are hidden during their first month of life, being visited by the female for suckling three of four times a day. At about one month old they will begin to accompany their mother during her daily foraging.

KEY FACTS

Size:	Height: 43 - 51 cms.
	Weight: 10 - 15 kgs (average).
Breeding:	Gestation: About 7 months.
	Young: One only.
	Sexual Maturity: One year old.
	Births: No specific season.
Lifespan:	Up to 15 years.
Lifestyle:	Family: Pairs or small family parties.
	Diet: Herbivorous.
	Main Predators: All major predators.
	Habitat: Rocky slopes, kopjes, cliffs and mountain sides.

Conservation & Status: Reasonably common in protected regions. In areas around human settlements they suffer constant competition from goat and sheep herds for available food.

Bushbuck

Order	Family	Genus & Species
Artiodactyla	*Bovidae*	*Tragelaphus scriptus*

Identification - One of the most elegant of Africa's antelopes, with rounded hindquarters slightly higher than the shoulders. The colour of the coat varies greatly from yellowish-chestnut through reddish-brown to dark brown. The underparts of the male are black. They have a dorsal mane of longer hair running from the shoulder to the tail. The head is lighter in colour with a dark band extending along the muzzle and a white cheek patch. The body has white vertical stripes and spots, mainly on the hindquarters and the back. The tail is bushy, white underneath, with a black tip. The horns are almost straight, with a single spiral, varying in length from 30 - 57 cms. and only occur in the male.

Distribution & Habits - An animal of forest edges and dense thickets, usually living singly or in pairs, but occasionally in small family groups. They are extremely secretive, hiding away during daylight hours in dense cover, feeding mainly during the night. Their main food consists of leaves and shoots, but they will dig for roots and tubers and plunder vegetable gardens from time to time. They will also associate with troops of baboons and monkeys, feeding on the fallen fruit shaken from the trees by the primates as they feed. When alarmed by the approach of a predator, the bushbuck's usual response is to freeze, in the hope of being overlooked.

KEY FACTS

Size:	Height: 61 - 96 cms.
	Weight: Males 60 kgs (average).
	Females 34 kgs (average).
Breeding:	Gestation: 6 months.
	Young: One only.
	Sexual Maturity: About 1 year.
	Births: No specific season.
Lifespan:	12+ years.
Lifestyle:	Family: Mainly singly or in pairs.
	Diet: Herbivorous.
	Main Predator: Leopard.
	Habitat: Dense forest and woodland edge.

Conservation & Status: Widely distributed and quite common in suitable habitats.

Sitatunga

Order	Family	Genus & Species
Artiodactyla	*Bovidae*	*Tragelaphus spekei*

Identification - A antelope with a shaggy medium length coat of drab grey/brown fur, often faintly marked with vertical stripes along the back and on the hindquarters. Some faint spotting also occurs on the flanks. The head has white patches on the cheeks and on the muzzle below the eyes. There are two white patches on the front of the neck, one on the throat and the other lower down towards the chest. The horns, which are only found in the male, are long, with spiral twists and measure up to 90 cms in length. The hooves are specially adapted to support their weight in wet habitats, being long and splayed at the tips. The animal has a hunched appearance with hindquarters higher than forequarters. The female is smaller than the male and usually has a coat more chestnut in colour.

Distribution & Habits - An aquatic antelope of marshes and swamps. Very secretive and shy, usually staying concealed in reeds and dense aquatic vegetation, only coming to the edge of cover in the evening. They feed mainly at dawn and dusk, as well as through the night, when they venture into surrounding grassland and wooded areas. They spend much of their time partially submerged and are very good swimmers. At the approach of danger they will often immerse themselves completely, all but for the tip of the nose. They often create well-trodden platforms of reeds and vegetation, on which to rest, by and action of trampling and turning. They also maintain a network of paths through the reedbeds. Local hunters often make use of these well defined pathways along which they place wire snares.

KEY FACTS

Size:	Height: 75 - 125 cms.
	Weight: Males 100 kgs (average).
	Females 54 kgs (average).
Breeding:	Gestation: Around 7 months.
	Young: One, very occasionally twins.
	Sexual Maturity: Males 1.5 - 2 years,
	Females 1 - 1.5 years.
	Births: No specific season. Females produce at yearly intervals.
Lifespan:	Up to 18 years.
Lifestyle:	Family: Singly or pairs.
	Diet: Herbivorous.
	Main Predators: Pythons, Lions, Leopards.
	Habitat: Swamps and marshes.

Conservation & Status: Under pressure from native hunters in some areas.

Lesser Kudu

Order	Family	Genus & Species
Artiodactyla	*Bovidae*	*Tragelaphus imberbis*

Identification - An antelope of slender build and medium size. The coat is grey/brown in colour and the sides of the body, from shoulders to rump, are marked with a series of light vertical stripes. The males carry large horns with 2 - 3 spirals, 60 - 90 cms in length. There are white patches on the cheeks, across the muzzle below the eyes and on the throat and lower neck. The long legs are tawny brown with patches of black and white. A line of short hair extends from the shoulders down the centre of the back to the tail, which is tawny/grey above, white below and black tipped. Females tend to be redder than males, while generally, the males become darker and greyer with age.

Distribution & Habits - An inhabitant of semi-desert regions of thorn bush country, seldom venturing into open areas. They feed by browsing bushes and trees for leaves, shoots, seedpods and fruits, occasionally taking fresh green grass. They are mainly active in the early morning, late afternoon and evening, during much of the day they will rest, either standing or lying, in dense thickets, where the broken pattern of the coat renders them almost invisible. When alarmed they will utter a loud bark and, when fleeing from predators, they are capable of leaping bushes and thickets up to 2 metres in height. Although they may visit waterholes for drinking when water is available, they are capable of surviving during long periods of dry weather, by obtaining sufficient liquid from their food.

KEY FACTS

Size:	Height: 100 - 110 cms.
	Weight: Males 90 kgs (average).
	Females 58 kgs (average).
Breeding:	Gestation: 7 months.
	Young: One only.
	Sexual Maturity: 18 - 24 months.
	Births: No specific season known. Females will calve on average every 8 - 10 months.
Lifespan:	Up to 15 years.
Lifestyle:	Family: Singly, in pairs or small groups.
	Diet: Herbivorous.
	Main Predators: Lion & leopard.
	Habitat: Dry thickets, thornbush country.

Conservation & Status: Although numbers have never been high they are probably stable. As the human demand for agricultural land and grazing for domestic stock increases they are likely to come under severe pressure.

Greater Kudu

Order	Family	Genus & Species
Artiodactyla	*Bovidae*	*Tragelaphus strepsiceros*

Identification - A very large elegant antelope, the males having magnificent horns with two and a half to three spirals and measuring up to 180 cms. in length. The smooth coat is bluish-grey to fawn with the sides of the body boldly marked with white vertical stripes, varying in number from 6 - 10. The head has a dark muzzle with white upper lip and chin and a conspicuous white stripe running from eye to eye across the bridge of the muzzle. A growth of longer hair extends down the centre of the back from the neck to the tail. The males have long hair growing along the throat and down the neck to the chest. The tail is of medium length, grey above, white below and with a black tip. The female is of much smaller build, very occasionally having horns.

Distribution & Habits - Greater Kudu inhabit thickets and areas of bush and scrub, often in hilly country and usually close to water, as they prefer to drink frequently if water is available, they can, however, survive in dry areas obtaining sufficient moisture from their food. They are usually found in small family herds of 4 or 5, but herds of 30 or more have been recorded. They avoid the heat of the day, browsing mainly in the early morning, late afternoon and into the night, on leaves, shoots and occasionally grasses. Females will isolate themselves to give birth, keeping the newborn hidden for 4 to 5 weeks, from then on the young will begin to accompany it's mother more and more. The young grow quickly and by about 6 months are fairly self reliant. Kudu are able to leap to heights of 2.5 metres to escape the pursuit of predators.

KEY FACTS

Size: Height: 100 - 150 cms.
Weight: Males 255 kgs (average).
Females 170 kgs (average).

Breeding: Gestation: About 7 months.
Young: One only.
Sexual Maturity: Males 5 years,
Females 2 years.
Births: Peak mating time in early dry season,
births peak during rains.

Lifespan: Up to 15 years.

Lifestyle: Family: Small family herds of 4 or 5.
Diet: Herbivorous.
Main Predators: Lion, Leopard, Hunting Dog.
Habitat: Thickets, bushes & scrub.

Conservation & Status: They are a species much prized by hunters, their horns are used by native tribes for a multitude of purposes, i.e. as containers and musical instruments.

Bongo

Order	Family	Genus & Species
Artiodactyla	*Bovidae*	*Tragelaphus euryceros*

Identification - A very striking antelope with a bright chestnut upper coat and dark almost black underside. The body from shoulder to rump is conspicuously marked with yellow-white vertical stripes, usually 12 to 14 in number. There is a crescent shaped stripe across the base of the neck. The ears are outlined with white hair and on the cheeks there are 2 or 3 white patches. There is also a white stripe on the bridge of the muzzle in front of the eyes. The legs have a broken pattern of irregular black, white and chestnut patches. The horns are large, ranging from 60 - 100 cms. in length, have a single spiral and vary in colour from light grey to black. They are less well developed in the female. The white tail is of medium length, terminating in a tuft of black hairs.

Distribution & Habits - An antelope of the densest mountain forests. Being very secretive and mainly nocturnal in habits makes this species one of the most difficult of the large mammals to see. They feed usually in dense cover, on leaves, shoots and the flowers of trees and bushes, occasionally grasses and herbs as well as roots which they unearth with their horns. They will sometimes stand on hind legs to reach higher foliage, using their long tongue to grasp bunches of leaves. They readily use saltlicks and require a constant supply of water throughout the year. The Aberdare Mountains of Kenya remains one of the best areas to see bongo. During the dry season they will move higher into the mountains, (2500 metres or higher), usually dispersing to lower elevations during the wet season.

KEY FACTS

Size:	Height: 120 -128 cms.
	Weight: Males 300 kgs (average).
	Females 240 kgs (average).
Breeding:	Gestation: 9.5 months.
	Young: One only.
	Sexual Maturity: 1.5 - 2 years.
	Births: No specific season known. Females will produce a calf annually.
Lifespan:	Up to 19 years.
Lifestyle:	Family: Singly, in pairs or small parties of females with young.
	Diet: Leaves, shoots, grasses and herbs.
	Main Predator: Leopard.
	Habitat: Dense montane forests.

Conservation & Status: A rare and endangered antelope.

Eland

Order	Family	Genus & Species
Artiodactyla	*Bovidae*	*Taurotragus oryx*

Identification - The largest African antelope, cattle-like in appearance. They have a hump on the shoulders and a dewlap at the base of the neck which generally becomes more pronounced in older males. General coloration is tawny-fawn to grey, the sides of the body being faintly marked with light vertical stripes, bolder on the shoulder than at the rear. Males have a crest of hair on the forehead and a mane on the nape. Horns are present in both sexes. They have several tight twists and range in length from 60 - 100 cms., generally thinner but longer in the female, sloping backwards following the profile line of the forehead. The tail is long, terminating in a tuft of black hair.

Distribution & Habits - An antelope of open plains and lightly wooded areas, thick forests being avoided. Gregarious by nature, herd sizes can vary from a few individuals to 500 or more. They feed on a variety of vegetation, taking large quantities of fresh grass following the rains, and browsing trees and bushes for leaves, shoots, fruits and seedpods during the dry season. They are active at all times of the day and night, but in extremely hot weather they will seek shade during the midday period. A newborn calf is able to walk almost immediately following its birth but will remain hidden for a period of around 2 weeks, before joining a nursery group along with it's mother and the rest of the herd. They do not generally allow a very close approach, moving away when vehicles are still several hundred yards distant.

KEY FACTS

Size:	Height: 125 - 175 cms.
	Weight: Males 700 kgs (average).
	Females 450 kgs (average).
Breeding:	Gestation: 9 months.
	Young: One only.
	Sexual Maturity: Males 4 - 5 years,
	Females 2.5 years.
	Births: No specific season, but peak births
	at end of dry season.
Lifespan:	Up to 20 years.
Lifestyle:	Family: Gregarious, herds of 15 - 20 average.
	Diet: Herbivorous.
	Main Predators: Lion and Spotted Hyaena.
	Habitat: Mainly open plains.

Conservation & Status: Relatively well distributed but under increasing threat from herds of domestic cattle and goats for available grassland.

Beisa Oryx

Order	Family	Genus & Species
Artiodactyla	*Bovidae*	*Oryx gazella beisa*

Identification - A large elegant and very distinctive antelope, with a short coat of grey-fawn, a black horizontal stripe running across the lower flank and black bands around the forelegs just above the knee joints. The facial markings are very conspicuous, a broad black stripe runs down the side of the face through the eye, from the base of the extremely long, thin horns to the lower cheek. The forehead has a black patch as does the upper muzzle. The hair on the lower muzzle is white, contrasting strongly with the black nostrils. A black line extends under the chin and around the upper throat from ear to ear. A thin black stripe extends down the centre of the neck from the throat to the chest and from the back of the neck along the spine to the root of the tail. The tail is long and horse-like, consisting of black hair. Another race of oryx, the Fringe-eared (Oryx gazella callotis) occurs in some regions and can be distinguished by a browner coat, slightly heavier build and tufts of long black hair on the points of the ears.

Distribution & Habits - A true desert species inhabiting dry arid regions with sparse bush cover and open areas of savannah. They feed mainly on grasses but will also browse trees and bushes. When water is available they will drink daily but they are capable of surviving for long periods without drinking. They are active throughout the day but generally seek shade during the midday period. They are to be found in herds numbering from 6 - 40 or more. Females seek isolation from the main herd in order to give birth, returning 2 to 3 weeks later along with the newborn. When under threat from predators they will usually flee, but will sometimes present a spirited defense, lowering the head they often inflict serious and sometimes fatal injuries upon their attacker with their spear-like horns.

KEY FACTS

Size:	Height: 120 cms.
	Weight: Males 175 kgs (average).
	Females 160 kgs (average).
Breeding:	Gestation: 8.5 - 10 months.
	Young: One usually, twins rare.
	Sexual Maturity: Males 5 years,
	Females 1.5 - 2 years.
	Births: No specific season.
Lifespan:	Up to 18 years.
Lifestyle:	Family: Herds of 6 - 40 average.
	Diet: Herbivorous.
	Main Predators: Lion, Leopard, Spotted Hyaena
	and Hunting Dog.
	Habitat: Arid regions.

Conservation & Status: Relatively common but under threat from the increase in domestic stock numbers in recent years.

Roan Antelope

Order	Family	Genus & Species
Artiodactyla	*Bovidae*	*Hippotragus equinus*

Identification - A large antelope with sloping shoulders and powerful neck and forequarters. The coat coloration varies from dark rufous to reddish-fawn contrasting with the black and white head and face markings. The ears are long and narrow with tufts of long hair at the tips. The under parts are white. A well defined mane of dark hair extends from the upper neck to the shoulders, a similar growth of hair is present on the underside of the neck extending down the centre of the throat. The legs are rufous, the forelegs having irregular black patches. The horns are heavily ringed and curve backwards in a sickle-shape, the length varies from 55 - 100 cms, they are less well formed in the female.

Distribution & Habits - An antelope of open and lightly wooded regions, living in small herds of about 20 animals, but they will gather together in larger herds when food and water are scarce, herds of 150 have been recorded. Each herd is led by a dominant bull. They feed mainly on grasses, occasionally browsing leaves and shoots from trees and bushes. They are very dependent on water and rarely venture far from a readily available source, they will usually visit water twice a day, taking large quantities. They are active throughout the day, usually finding cover and shade during the hottest period around midday. Females calve away from the herd and hide the newborn in dense vegetation for some weeks, only returning to suckle the infant twice a day.

KEY FACTS

Size:	Height: 125 - 160 cms.
	Weight: 270 kgs (average).
Breeding:	Gestation: 9 - 9.5 months.
	Young: One only.
	Sexual Maturity: 2.5 - 3 years.
	Births: No specific season, females produce a calf on average every 10 - 11 months.
Lifespan:	14+ years.
Lifestyle:	Family: Herds of 20 on average.
	Diet: Herbivorous.
	Main Predators: Lion, & Leopard. Calves are vulnerable to Spotted Hyaena and Hunting Dogs
	Habitat: Open, lightly wooded country.

Conservation & Status: Not at all numerous, on the decline due to the conversion to agriculture of suitable habitat.

Sable Antelope

Order	Family	Genus & Species
Artiodactyla	*Bovidae*	*Hippotragus niger*

Identification - A large and magnificent antelope with powerful forequarters. The male has a glossy black coat, white underparts and a striking head of black and white markings. White hair covers the chin and the tip of the muzzle, from where broad white bands extend up the sides of the face terminating on the forehead. The crown of the head and the outside of the ears are light chestnut, the inside of the ears are white. They have a pronounced mane of stiff hairs from the neck to the shoulders and a long black tail. Females and young are often paler and more chestnut in colour than the males. The horns are very long, measuring up to 154 cms in length, sweep backwards in an arc and are heavily ridged, they are less well developed in the female.

Distribution & Habits - An inhabitant of light woodland, bush and grasslands. They are found in herds of 10 - 20 animals on average but will occasionally form larger herds, particularly during the dry seasons when suitable grazing is much reduced. The herds, of mainly females and young, are usually headed by a dominant bull and fights with neighbouring males are quite frequent, the combatants dropping to their knees and engaging horns in bouts of head wrestling. Fatalities are known but they are very rare. They are primarily grazers taking grasses and herbs but occasionally they browse trees and bushes for leaves and shoots. They are heavily dependent on water and although they only drink on average every other day, they rarely wander too far from a dependable source. They are active mainly in the early morning and late afternoon and evening.

KEY FACTS

Size	Height: 117 - 140 cms.
	Weight: 220 - 235 kgs (average).
Breeding	Gestation: 8 - 9 months.
	Young: One only.
	Sexual Maturity: 2.5 - 3 years.
	Births: Peak in Jan/Feb.(Kenya)
Lifespan:	16+ years.
Lifestyle	Family: Herds of 10 - 20 on average.
	Diet: Herbivorous.
	Main Predators: Lion, Leopard,
	Spotted Hyaena.
	Habitat: Lightly wooded regions.

Conservation & Status: Distribution has been much reduced in recent years as a result of the spread of human settlements and, consequently a demand from domestic animals for food and water.

Common Waterbuck

Order	Family	Genus & Species
Artiodactyla	*Bovidae*	*Kobus ellipsiprymnus*

Identification - A large antelope with a shaggy coat of grizzled grey with an occasional brown tinge, sometimes the back and lower legs are darker, becoming almost black. The white ears are large, with black tips. Facial markings are limited to a white stripe extending from the eyebrow along the sides of the muzzle to just below the eye, and a white patch around the nasal area and the lips. They have a white collar extending across the throat, almost from ear to ear. On the rump is a pronounced white crescent shape, which distinguishes it from the sub-species Defassa Waterbuck (Kobus e. defassa), which has a circular white patch on the buttocks. The tail is of medium length having a dark tip. The horns, present only in the male, are heavily ringed and curve backwards, upwards and forwards towards the tips.

Distribution & Habits - Waterbuck inhabit woodlands and clearings usually close to water, where they feed mainly on grasses and herbs. They will also occasionally feed on the foliage of trees and bushes, particularly during periods of extreme drought. They are dependent on water and will drink every day or so. Mainly active during the early morning, late afternoon and early evening. They are usually found in small groups consisting of a bull with several females and young numbering on average 5 - 10 animals in all. Herds of bachelor males may be encountered of 6 - 40 in number, males are at least 4 years old before they are able to establish themselves as master bulls. When pursued by predators waterbuck will flee and attempt to hide themselves in bush cover or long grass, but there have been many reports of them taking refuge in water, submerging completely all but for the nostrils.

KEY FACTS

Size	Height: 120 - 136 cms.
	Weight: Males 240 kgs (average).
	Females 180 kgs (average).
Breeding	Gestation: 8.5 months.
	Young: One usually, twins rare.
	Sexual Maturity: 12 - 18 months.
	Births: No specific season. Females produce on average every 10 - 11 months.
Lifespan:	16+ years.
Lifestyle	Family: Small family herds.
	Diet: Herbivorous.
	Main Predators: Lion, Leopard, Spotted Hyaena.
	Habitat: Woodlands and clearings.

Conservation & Status: Relatively common in suitable habitats but under constant pressure from domestic animals for food and water.

Defassa Waterbuck
A sub-species of Common Waterbuck

Order	Family	Genus & Species
Artiodactyla	*Bovidae*	*Kobus ellipsiprymnus defassa*

Identification - A large antelope with a long, coarse shaggy coat, the Defassa Waterbuck is a sub-species of the Common Waterbuck and is very similar in appearance. The main visual difference being the pattern on the rump, which in Defassa is a solid white patch radiating from the base of the tail and covering the buttocks, while in the Common Waterbuck the rump markings take the form of white semi-circular stripes. For other identification details see Common Waterbuck. In areas where the distribution of the two overlaps, inter-breeding has been recorded resulting in intermediate rump patterns.

Distribution & Habits - The habitat preference of Defassa Waterbuck is very similar to that of the Common Waterbuck, woodlands and clearing in the vicinity of water, where they feed on grasses and browse leaves and shoots from trees and bushes. The Defassa has a greater and wider distribution throughout Africa than that of the Common and seems to prefer areas with a pattern of greater rainfall. Both have coats impregnated with oil which is secreted from sweat glands, which may serve as water-proofing as well as for individual recognition. They form small family herds of 5 - 10 animals as well as bachelor herds of up to 40.

KEY FACTS

Size	Height: 120 -136 cms.
	Weight: Males 240 kgs (average).
	Females 180 kgs (average).
Breeding	Gestation: 8.5 months.
	Young: One usually, twins are rare.
	Sexual Maturity: 12 - 18 months.
	Births: No specific season, some inter-breeding with Common Waterbuck.
Lifespan:	16+ years.
Lifestyle	Family: Small family herds.
	Diet: Herbivorous.
	Main Predators: Lion, Leopard and Spotted Hyaena.
	Habitat: Woodlands and clearings.

Conservation & Status: Relatively common in many suitable areas but under increasing threat from the spread of human settlements and the competition from domestic animals for food and water.

Uganda Kob

Order	Family	Genus & Species
Artiodactyla	*Bovidae*	*Kobus kob*

Identification - A graceful antelope of medium size with a short coat of light cinnamon to brown. The underparts are white and the forelegs have a black stripe on the front above and below the knee joints. A light, almost white, area of hair encircles the eyes and a white patch appears on the underside of the throat and upper neck. The insides of the ears are white and the tail is of medium length terminating in a tuft of black hair. The lyre-shaped horns are only present in the male, they are deeply ringed and curve backward, outward then straighten before arcing upward to form an 'S' in profile. They occasionally reach 70 cms. in length.

Distribution & Habits - An animal of open grassy savannah and lightly wooded areas in the west of the region. They generally live in herds of 20 to 40 animals. They feed on grasses and herbs, as well as occasionally grazing on aquatic vegetation, they are rarely found far from water. They are active throughout the day even on occasions grazing during the hottest midday period. Males are generally 3 to 4 years old before successfully securing and holding a territory, the females moving at will from one territory to another in search of the strongest males with which to breed.

KEY FACTS

Size	Height: 82 -100 cms.
	Weight: Males 93 kgs (average).
	Females 62 kgs (average).
Breeding	Gestation: 8 -9 months.
	Young: One only.
	Sexual Maturity: 1.3 - 1.5 years.
	Births: No specific season.
Lifespan:	Up to 16 years.
Lifestyle	Family: Herds of 20 - 40 average.
	Diet: Herbivorous.
	Main Predators: Lion, Leopard, and Spotted Hyaena.
	Habitat: Open savannah and sparse woodland.

Conservation & Status: Relatively common in suitable habitats.

Mountain Reedbuck

Order	Family	Genus & Species
Artiodactyla	*Bovidae*	*Redunca fulvorufula*

Identification - A small antelope with a shaggy coat of grey-fawn, the head and neck are tinged with rufous-brown, the chest and belly are white. The eyebrows, throat, chin and lips are faintly marked with off-white as is the short, bushy tail. A gland of dark skin is clearly visible below the ears on the sides of the head, a dark line extends down the centre of the muzzle from the eyes to the nose. Horns are only present in the male and are short, heavily ringed and arc forwards. Females are usually larger and often greyer in colour than the males.

Distribution & Habits - An antelope of grassy hill sides and mountains from altitudes of 1500 metres upwards, where they feed by grazing grasses and herbage. Although drinking regularly when water is available, they are able to obtain sufficient moisture from their food intake to survive prolonged periods of drought. They gather in small herds of 3 - 8 animals, usually consisting of females and young occupying the territory of a single resident male. Other males tend to form small bachelor groups. They are active in early morning and late afternoon as well as during moonlit nights. Females are able to produce young from 1 year old and, on average, will produce a calf every 9 - 14 months. Females having given birth will leave the young hidden for a period of 2 - 3 months, visiting only for short periods of suckling.

KEY FACTS

Size	Height: 64 - 76 cms.
	Weight: 30 kgs (average).
Breeding	Gestation: 8 months.
	Young: One only.
	Sexual Maturity: Females from 12 months.
	Births: No specific season.
Lifespan:	Up to 12 years.
Lifestyle	Family: Small herds of 3 - 8 animals.
	Diet: Herbivorous.
	Main Predators: Lion, Leopard, Cheetah and Spotted Hyaena.
	Habitat: Grassy hill sides and mountains.

Conservation & Status: Very much under threat outside National Parks.

Bohor Reedbuck

Order	Family	Genus & Species
Artiodactyla	*Bovidae*	*Redunca redunca*

Identification - An antelope of medium size, having a thick coat of uniform reddish-fawn. The underside of the body is whitish-grey. The same white-grey fur appears around the eyes, on the cheeks, lips, chin, throat and on the insides of the large oval ears. The nose and the centre of the upper lip are black. Occasionally a black stripe is discernable extending down the front of the forelegs. The tail is short with a bushy tip, the white underside being very conspicuous when exposed by the animal when fleeing in alarm. The horns, present only in the male are lyre-shaped, curving forwards and inwards and are heavily ringed particularly towards the base.

Distribution & Habits - An animal of open plains, lush wet grasslands and swampy ground usually in the vicinity of water. They are usually found in pairs or in small family herds of 3 to 6 animals and feed almost exclusively on grasses. They are active throughout the day, as well as during clear moonlit nights, however, they usually seek the cover of long grass or bushes during the hot midday period, sitting very tight if approached and bolting only at the last moment. Sightings of very young animals are extremely rare, the females keeping them very well hidden for a period of at least two months.

KEY FACTS

Size	Height: 70 - 90 cms.
	Weight: Males 43 kgs (average).
	Females 36 kgs (average).
Breeding	Gestation: 7 - 7.5 months.
	Young: One only, twins are rare.
	Sexual Maturity: About 18 months.
	Births: No specific season.
Lifespan:	Up to 10 years.
Lifestyle	Family: Pairs or small parties.
	Diet: Herbivorous.
	Main Predators: All major predators as well as Jackals, Eagles and Pythons which prey on young.
	Habitat: Grasslands, swamps.

Conservation & Status: Relatively common in suitable habitats.

Topi

Order	Family	Genus & Species
Artiodactyla	*Bovidae*	*Damaliscus lunatus*

Identification - A large antelope with a pronounced 'slope', being higher in the forequarters than at the rear. The coat is short and glossy, dark chestnut-brown with bluish-black patches extending from the lower shoulders down the forelegs to just above the knees and from the rump down the back legs terminating just above the hock. The same bluish-black hair extends down the centre of the long narrow face, from the forehead to the tip of the muzzle and on the backs of the ears. The horns, 30 to 60 cms. in length, appear in both sexes but are larger in the male. They are heavily ringed and angle backwards, slightly outwards and upwards towards the tips. The tail is of medium length, terminating with tufts of long black hair.

Distribution & Habits - An antelope of open grassland and areas of light scattered bush and scrub, where they feed almost entirely on grasses and herbage. They are found in small groups of around 12 animals, usually consisting of a dominant bull along with a harem of females and young. They can, however, congregate in herds of thousands during the dry season when grazing is scarce. They will readily mingle with Hartebeest, Wildebeest, Zebra and other plains ungulates. Although they will drink regularly when water is available, they are able to survive up to a month without water. They are active both by day and by night. An individual can often be found standing on a termite mound, or similar piece of elevated ground, on the lookout for predators, whilst other members of the herd continue to graze.

KEY FACTS

Size	Height: 108 - 134 cms.
	Weight: 130 - 140 kgs (average).
Breeding	Gestation: 7.5 - 8 months.
	Young: One only.
	Sexual Maturity: Males 3 years,
	Females 1.5 - 2 years.
	Births: Seasonal peak in Feb/March in certain areas.
Lifespan:	Up to 15 years.
Lifestyle	Family: Usually small family herds of around 12 animals.
	Diet: Herbivorous.
	Main Predators: Lion, Leopard, Cheetah and Spotted Hyaena.
	Habitat: Open plains and lightly wooded regions.

Conservation & Status: Numbers greatly reduced in recent years due to hunting and habitat loss.

Cokes Hartebeest

Order	Family	Genus & Species
Artiodactyla	*Bovidae*	*Alcelaphus buselaphus*

Identification - A large antelope, characterised by it's 'sloping' appearance, being higher in the forequarters than in the rear. The coat is of uniform yellow-fawn, lighter on the hindquarters with an almost white rump and underside. The head and face are long and narrow with large pointed ears. The horns are sickle-shaped, curving forwards, outwards then backwards, are deeply ringed and are present in both sexes. The tail is of medium length, terminating in a tassel of black hair. The female is similar to the male but usually paler in colour and with smaller, slimmer horns.

Distribution & Habits - One of the most common of Africa's antelopes, inhabiting areas of open grassland and dry savannah, sometimes with lightly scattered trees and bushes. They can be found in herds ranging in number from just a few to many thousands, the larger concentrations occurring during the dry season when available grazing is limited. Old males may be solitary and young males may form into bachelor herds. They are active mainly in the early morning and late afternoon and evening, seeking shelter from the sun during the hot midday period. They feed mainly on grasses and herbage and will drink every day when water is available, however, during prolonged periods of drought they can survive on the moisture obtained from their food. They often associate with other species of plains herbivores.

KEY FACTS

Size	Height: 107 - 120 cms.
	Weight: 125 - 140 kgs (average).
Breeding	Gestation: 8 months
	Young: One only, twins rare.
	Sexual Maturity: 2 - 2.5 years.
	Births: No specific season, small peak in Jan/March period.
Lifespan:	Up to 18 years.
Lifestyle	Family: Herds of varying sizes.
	Diet: Herbivorous.
	Main Predators: Lion, Leopard and Spotted Hyaena, other small predators i.e. jackals prey on young calves.
	Habitat: Open grassland and lightly wooded regions.

Conservation & Status: Has shown a marked decline in recent years due to hunting and the inability to compete with domestic stock for suitable grazing.

Wildebeest

Order	Family	Genus & Species
Artiodactyla	*Bovidae*	*Connochaetes taurinus*

Identification - A large antelope with forequarters higher than hindquarters giving the appearance of a pronounced 'slope' from front to rear. The coat is greyish in colour, with a long mane of black hair on the neck and shoulders and a long beard of hair on the throat. The coloration of the beard varies greatly from almost pure white to black. The neck, shoulders and, to a lesser extent, the flanks show dark vertical stripes. The head is large and broad with a completely black face. The horns are present in both sexes, measuring between 40 to 73 cms. in length and from a flat base curve outwards, downwards then upwards, not dissimilar in shape to those of the Cape Buffalo. The black bushy-tipped tail is extremely long almost touching the ground.

Distribution & Habits - An extremely common antelope of open grassland and lightly wooded regions where they gather in enormous herds during their annual migration in a continual search for fresh grazing. This circular, seasonal migration from the southern Serengeti to the Masai Mara and back covers a distance in excess of 800 kms. and at times columns of animals stretch for 40 kms. They eat mainly grass and are water dependent drinking twice a day if possible. They are active throughout the day usually attempting to rest or seek shade during the hot midday period. The females synchronise their calving with 80% of young being born during a three week period usually towards the beginning of the rainy season (Feb - March). To avoid predation the females and newborn need to remain mobile and to this end the calves are on their feet within 3 to 7 minutes of birth. This synchronised calving produces an abundance of food for predators, but over a short period, thereby ensuring the survival of the majority.

KEY FACTS

Size	Height: 125 - 140 cms.
	Weight: 160 -200 kgs (average).
Breeding	Gestation: 8.5 months.
	Young: One only.
	Sexual Maturity: Males 4 - 5 years,
	Females 1.5 - 2 years.
	Births: Peak during Feb/March.
Lifespan:	Up to 20 years.
Lifestyle	Family: Large herds.
	Diet: Herbivorous.
	Main Predators: Lion, Leopard, Cheetah,
	Spotted Hyaena.
	Habitat: Open grassland and lightly wooded savannah.

Conservation & Status: Numbers have increased during the past few decades, the main factor is thought to be the absence of any major epidemic such as rinderpest.

Impala

Order	Family	Genus & Species
Artiodactyla	*Bovidae*	*Aepyceros melampus*

Identification - A very graceful medium sized antelope with a long neck, a smooth short coat of rufous-brown, being paler on the lower flanks and with white underparts. The males carry magnificent, wide-set, lyre-shaped horns that curve backwards, sideways then upwards, they are heavily ringed and can grow to over 90 cms. in length. The upper lip, chin, throat and eyebrows are white. The inner ears are white tipped with black. The nose is black and forms a characteristic 'Y' shape. A vertical black line extends down the hindquarters from the root of the tail. Just above the heel of the hind legs there is a tuft of long black hair. The tail is of medium length with a black stripe on the upper side. The female is smaller than the male and lacks horns.

Distribution & Habits - A species of open plains and sparse woodlands, within easy reach of water, they will drink twice a day when water is available. They feed mainly on short grasses which forms up to 95% of their food intake, only occasionally browsing from trees and bushes. They are normally encountered in small herds of around 6 to 20 animals, although herds of 50+ are quite common. The herds usually consist of a single dominant male and females with young. Males are usually 4 to 6 years old before attaining the strength and stature to hold a territory and secure a harem of females, fights between rival males during the rutting season can be very fierce. During the dry season they often form larger herds, concentrated together on the small areas of remaining grassland. Their sight is not well developed but their hearing and smell are acute. When under threat from predators they are quite fast and capable of leaps of 11 metres in length and 3 metres in height.

KEY FACTS

Size	Height: 70 -92 cms.
	Weight: 45 - 60 kgs (average).
Breeding	Gestation: 6.5 - 7 months
	Young: One only.
	Sexual Maturity: 18 months.
	Births: No specific season, slight peaks during Aug - Nov and Mar - May.
Lifespan:	Up to 12 years.
Lifestyle	Family: Herds of 6 - 20 on average.
	Diet: Herbivorous.
	Main Predators: All major predators. Eagles will take small young.
	Habitat: Open plains and light woodlands.

Conservation & Status: Numbers seem to be steady at the moment, remaining numerous in suitable areas.

Grants Gazelle

Order	Family	Genus & Species
Artiodactyla	*Bovidae*	*Gazella granti*

Identification - A large elegant antelope with long legs and powerful build. The short, smooth coat is fawn often with the head and neck paler. The insides of the legs and the underparts are white. The buttocks are white contrasting with a dark patch extending down the rump. The white of the buttocks continues to a point just above the root of the tail and curves out for a short distance on to the rump. In younger animals the point along the lower flank where the fawn coat meets the white underside is often marked with a darker horizontal band. The head is marked with white patches around the eyes, along the sides of the muzzle, around the nasal area and the chin. A black line extends from the eyes down the muzzle to the corners of the mouth. A black band sweeps over the muzzle just above the nasal area. The horns are up to 80 cms. in length, heavily ringed and lyre-shaped curving backwards, outwards and upwards, the outward divergence is very variable.

Distribution & Habits - An inhabitant of open plains and dry bush country. Males of at least 3 years old establish territories and gather together females forming harems of 10 - 30 in number, larger herds form during non-rutting seasons. They feed on grasses and leaves and do not normally need to drink, obtaining sufficient liquid from their food, this allows them to exploit feeding opportunities in drier regions which water dependent herbivores are unable to inhabit. There is some evidence of seasonal migration in search of grazing. They graze mainly during the early morning and late afternoon but remain active throughout the day. Females will leave the herd to calve, moving to areas offering cover, in which the newborn can hide for the first 4 - 6 weeks of life.

KEY FACTS

Size	Height: 75 - 91 cms.
	Weight: 45 -65 kgs (average).
Breeding	Gestation: 6.5 months.
	Young: One only.
	Sexual Maturity: 18 months.
	Births: No specific season.
Lifespan:	10 -12 years.
Lifestyle	Family: Herds of 10 -30 on average.
	Diet: Herbivorous.
	Main Predators: Most large predators.
	Habitat: Open plains and dry bush.

Conservation & Status: Numbers appear to have increased in recent years.

Thomson's Gazelle

Order	Family	Genus & Species
Artiodactyla	*Bovidae*	*Gazella thomsoni*

Identification - Medium-small in size this delicate gazelle is sandy-rufous in colour, darker along the back and paler on the head, neck, lower flanks and rump. The chest, belly and insides of the legs are pure white. A very conspicuous thick black band extends horizontally from the top of the forelegs across the flanks to the thighs. The buttocks are white bordered by a thin black stripe. The forehead and centre of the muzzle are dark rufous-brown, broad bands of white hair extend from the base of the ears through the eyes to the tip of the muzzle. A black stripe extends from the eyes down the cheeks to the corners of the mouth. The lips, chin and throat are white. The horns are deeply ringed and quite long, up to 40 cms. or more, they are only slightly curved and are much reduced in the female. They have a short black tail.

Distribution & Habits - A common antelope of open plains and grasslands, named after the explorer Joseph Thomson who first saw the species in Kenya in 1883. They live in harem herds of one dominant male with, on average, 6 - 60 females, but during certain periods of the year, usually the dry season, they gather into herds of many thousands, mixing with other ungulates. They feed almost exclusively on grasses, only occasionally browsing from trees and bushes. They will drink daily when water is readily available but can survive for long periods without. They are most active during the early morning and late afternoon and evening. Newborn calves remain hidden between periods of suckling and in open country their coloration and the ability to 'freeze' makes them extremely difficult to detect. The young grow rapidly being weaned within two months.

KEY FACTS

Size	Height: 58 - 70 cms.
	Weight: Males 23 kgs (average).
	Females 18 kgs (average).
Breeding	Gestation: 6 months.
	Young: One only.
	Sexual Maturity: Males 2 years,
	Females 1 year.
	Births: No specific season.
Lifespan:	Up to 10 years.
Lifestyle	Family: Herds of 6 - 60 on average.
	Diet: Herbivorous.
	Main Predators: All predators including
	Eagles, Pythons and occasionally Baboons.
	Habitat: Open plains and grasslands.

Conservation & Status: Hunted in many areas for the 'pot', although generally any decline in numbers is thought to be slight.

Gerenuk

Order	Family	Genus & Species
Artiodactyla	*Bovidae*	*Litocranius walleri*

Identification - A large, slender built gazelle, almost unmistakable having an extremely long neck and long delicate legs. The upper body colour along the centre saddle of the back is reddish-buff, becoming lighter on the flanks, rump, legs, neck and head. The underside is white and clearly defined. A narrow vertical white stripe is present either side of the root of the tail. The tail itself is short and tipped with a tuft of black hair. The head is small and narrow, the large dark eyes are ringed with patches of white hair in which are set dark preorbital scent glands. The ears are large, the insides being white edged with prominent black markings. The horns are restricted to the males only and are short but substantial, measuring up to 44 cms. in length curving backwards, outwards and turning forwards and upwards towards the tips, they are heavily ringed. Females are slighter in build than males and show a dark patch on the crown of the head.

Distribution & Habits - Living singly, in pairs or small groups consisting of a dominant male with several females and young, the Gerenuk inhabits areas of dry bush and scrub. They feed exclusively on the foliage of trees and bushes. They often reach higher leaves by standing erect on their back legs, this action coupled with the elongated neck allows them to browse leaves and shoots that are out of the reach of most other animals, reaching heights in excess of two metres. They do not need to drink, obtaining all their liquid requirement from their food matter. They are active in the early morning and in the late afternoon and evening, resting in shade if possible during the hot midday period.

KEY FACTS

Size	Height: 80 -105 cms.
	Weight: 30 - 45 kgs (average).
Breeding	Gestation: About 7 months.
	Young: One only.
	Sexual Maturity: Males 18 months,
	Females 12 months.
	Births: No specific season.
Lifespan:	10 - 12 years.
Lifestyle	Family: Singly, in pairs or small family groups.
	Diet: Herbivorous.
	Main Predators: All major predators.
	Smaller predators i.e. Jackals will take young.
	Habitat: Dry bush country.

Conservation & Status: Relatively common in suitable habitats but constantly under threat from the spread of domestic stock.

Cape Buffalo

Order:	Family	Genus & Species
Artiodactyla	*Bovidae*	*Syncerus caffer*

Identification - An enormous bovid of strong, solid build with short legs and cattle-like appearance. The sparse coat is short and blackish in colour, young animals have a thicker browner coat. The stout muscular neck supports a large head with a wide muzzle and large ears on the sides of the head, beneath massive horns which spread outwards and downwards from a thick broad base before arcing upwards and inwards. The horn size and shape is variable and dependent on age, the old mature bulls carrying the prize sets, they are much reduced in the female. The tail is long terminating in a tassel of black hair.

Distribution & Habits - Found over a wide range of habitats from dense forests and woodlands to open plains. They are primarily grazers but in forest habitats will browse leaves and shoots. They are water dependent and need to drink daily if possible. They are gregarious, living in herds of 20 - 40 animals on average, but sometimes form into herds of several hundreds. Old bulls tend to live a solitary existence or form small geriatric herds. They are often attended by flocks of Oxpeckers or Tick Birds who do a service to the animals by removing ticks and blood-sucking insects from the hide. They are active by both day and night, seeking shelter during the hot midday period. They take great delight in wallowing in mud holes. The eyesight is rather poor and the hearing only average but the sense of smell is acute. In the face of danger or uncertainty they will usually lumber away, but they need to be treated with the utmost respect, being responsible for many serious injuries and deaths among local people.

KEY FACTS

Size	Height: 135 -170 cms.
	Weight: Males 680 kgs (average).
	Females 480 kgs (average).
Breeding	Gestation: 11.5 months.
	Young: One only, twins very rare.
	Sexual Maturity: Around 3 years.
	Births: All times of year with peak at beginning of the rains.
Lifespan:	Up to 20 years.
Lifestyle	Family: Herds of 20 -40 on average.
	Diet: Herbivorous.
	Main Predator: Lion.
	Habitat: Open plains to dense forest.

Conservation & Status: Their numbers have increased in recent years, they are, however, susceptible to rinderpest outbreaks which can greatly deplete numbers.

Grevy's Zebra

Order	Family	Genus & Species
Perissodactyla.	*Equidae*	*Equus grevyi*

Identification - The distinctive markings of the zebra make for easy identification. The Grevy's is the largest member of the zebra family with narrow black stripes, on a background of white, covering the body, head and the entire length of the legs to the hooves. The neck and body stripes are vertical, those on the neck continuing through the stiff hair of the mane, whilst those covering the legs are horizontal. The stripes terminate on the lower body leaving a plain white under-belly. A broad black stripe extends from the base of the mane to the root of the tail. Each animal has a pattern as unique as a finger print. They have large rounded ears, black edged and tipped with white. The point of the muzzle around the nose and mouth is black often tinged with brown. The tail is long, rounded and striped at the base, terminating in a tuft of long black hair. Young animals often have the dorsal mane and the rear and upper body stripes brownish.

Distribution & Habits - A zebra of dry, semi-desert regions, where they are normally found in small herds of 6 to 20 animals, although large herds of several hundreds do gather from time to time. Males often gather in bachelor herds, usually reaching full maturity around 6 years old. In northern Kenya, the last stronghold of the species, they often associate with herds of Beisa Oryx. They feed by grazing on grasses and herbage, being mainly active in the early morning and late afternoon. They are less dependent on water than other zebra species but drink daily when water is available and can only survive for a short period without liquid, often digging in river beds and stream bottoms to reach the subterranean water level.

KEY FACTS

Size	Height: 140 -160 cms.
	Weight: Males 430 kgs (average).
	Females 385 kgs (average).
Breeding	Gestation: Around 12 months.
	Young: One only.
	Sexual Maturity: 3.4 - 4 years.
	Births: No specific season, slight peak July/August.
Lifespan:	Up to 20 years.
Lifestyle	Family: Herds of 6 - 20.
	Diet: Herbivorous.
	Main Predators: Mainly Lion and Spotted Hyaena.
	Habitat: Dry, semi desert regions.

Conservation & Status: Endangered. Numbers have decreased in recent years as a result of hunting for skins and increased competition for grazing and water from domestic livestock.

Common Zebra

Order	Family	Genus & Species
Perissodactyla	*Equidae*	*Equus quagga*

Identification - Africa's version of the horse, the distinctive markings make the zebra easy to identify. Broad blackish stripes, vertical on the neck and shoulders and horizontal on the rump and legs are set against an almost white body colour. The stripes continue down the legs to the hooves, cover the head, face and the stiff hair of the mane, each animal having a pattern as unique as a finger print. The ears are large. The tip of the muzzle, around the nose and mouth, is black and sometimes tinged with brown. A broad black stripe follows the line of the spine from the base of the mane to the root of the tail. The stripes of young animals are usually brown rather than black and have a shaggy appearance.

Distribution & Habits - An animal of open plains, grasslands, hills and mountains, where they roam in large herds often numbering thousands, but generally they are found in smaller family groups of 6 to 20 animals consisting of a dominant stallion of at least 5 to 7 years old, with a number of mares and foals. They are primarily grazers but will occasionally browse leaves and shoots from a variety of trees and bushes. They are dependent on water and are rarely found far from a permanent source. They are active throughout the day. They are subject to seasonal migrations constantly searching for better grazing, often mixing with Wildebeest herds. At the approach of danger adults may present a collective defense for themselves and other herd members, particularly young.

KEY FACTS

Size	Height: 127 - 140 cms.
	Weight: 220 -250 kgs (average).
Breeding	Gestation: 12 months.
	Young: One only. Twins rare.
	Sexual Maturity: Males 2.5 - 3 years,
	Females 1.5 years.
	Births: Anytime but peak Dec/Feb
Lifespan:	Around 20 years.
Lifestyle	Family: Herds of 6 - 20 average.
	Diet: Herbivorous.
	Main Predators: Lion, Spotted Hyaena, Hunting Dog.
	Habitat: Varied. Open plains, hills and mountains.

Conservation & Status: Numbers much reduced, as a result of hunting for skins and the increased competition for grazing and water presented by domestic livestock.

Black Rhinoceros

Order:	Family	Genus & Species
Perissodactyla	*Rhinocerotidae*	*Diceros bicornis*

Identification - A relic of prehistoric times the rhinoceros is almost unmistakeable. The Black Rhino is distinguished from the slightly larger White Rhino, by the narrow mouth and prehensile upper lip. The head is large and carries two horns, the larger front horn measuring on average 60 cms., although individuals with a front horn over twice that length have been recorded. The ears are oval and tipped with tufts of dark hair. The eyes are small and the eyesight poor, but the senses of smell and hearing are very acute. The huge body is covered with a thick hide of grey skin, although due to the rhino's predilection for wallowing in mud, their coloration can appear very variable. The tail is short and tipped with stiff hairs. In spite of their bulk the Black Rhino is very manoeuvrable and capable of a top speed of 50 kph.

Distribution and Habits - The distribution of this species has been much reduced in the past 25 years as a result of increased poaching. They favour areas of dry bushy savannah and lightly wooded regions. In most areas they survive by browsing leaves and shoots from bushes and trees, although the population inhabiting the Ngorongoro Crater grazes regularly due to the lack of suitable browse. They feed mainly in the early morning and late afternoon, seeking shade or a mud wallow during the hottest midday period. Although they are capable of surviving for several days without water, they will drink and wallow daily when possible, often travelling many miles to an available source. During periods of drought they will often dig for water in dried up river beds using their forelegs. Rhino's are solitary animals although females are usually accompanied by their most recent offspring.

KEY FACTS

Size	Height: 1.6 m
	Weight: Up to 1400 kgs.
Breeding	Gestation: 16 months.
	Young: One only.
	Sexual Maturity: 6-7 years.
	Births: No specific season, but peaks after the rainy season.
Lifespan	Up to 40 years.
Lifestyle	Family: Solitary.
	Diet: Herbivorous.
	Main Predators: Man. Lions and Spotted Hyaenas will occasionally take an unguarded calf.
	Habitat: Dry, bushy savannah and light woodlands.

Conservation & Status: Numbers have been drastically reduced during the past 25 years as a result of poaching. Fuelled by an increase in the demand for horns by Arab nations for ceremonial dagger handles and by Far Eastern countries, as a medicine/aphrodisiac, the species is threatened with extinction.

White Rhinoceros

Order:	Family	Genus & Species
Perissodactyla	*Rhinocerotidae*	*Ceratotherium simum*

Identification - The White Rhino is the world's second largest land mammal - the Elephant being the largest. The animal's name has nothing whatever to do with colour, but is a corruption of the Afrikaans word "Weit" meaning 'wide' and refers to the shape of the mouth, this being the most obvious difference between the two African species. Far bigger and heavier than the Black Rhino the head is large, square-shaped and carries two horns. The front horn is the larger of the two, averaging 60 cms. in length, the rear horn is shorter and thicker. The ears are large and oval in shape, the eyes are small and the eyesight is rather poor. The huge body is covered with a thick hide of pale grey skin. The tail is short, terminating with stiff hairs.

Distribution & Habits - Several attempts to reintroduce this species into East Africa's National Parks have been undertaken in recent times, but most of the introduced animals have subsequently fallen to poachers guns and snares. Hopefully, the latest introduction of seventeen animals from South Africa into Kenya's Lake Nakuru National Park, will fare better, thereby securing the long-term future of the species in East Africa. They feed by grazing, using the wide mouth and strong lips to crop short grasses. They will feed throughout the day and night, but usually seek shelter from the hot midday sun. They will drink several times a day if water is readily available but during the dry season they can subsist by drinking every 3 to 4 days. In spite of their bulk they are suprisingly quick and manoeuvrable.

KEY FACTS

Size:	Height: 1.85 m.
	Weight: 2100 kgs (average)
Breeding	Gestation: 16 months.
	Young: One only.
	Sexual Maturity: 6-7 years.
	Births: No specific season, but peak occurs after the rains.
Lifespan	Up to 40 years.
Lifestyle	Family: Solitary, although from time to time animals, usually of the same sex, may gather together.
	Diet: Herbivorous.
	Main Predator: Man.
	Habitat: Dry, bushy savannah & lightly wooded regions.

Conservation & Status - Like the Black Rhino the long-term survival of this species in the wild is in some doubt. The persistent and systematic poaching of this most gentle of animals, is bringing it to the verge of extinction.

Tree Hyrax

Order	Family	Genus & Species
Hyracoidea	*Procaviidae*	*Dendrohyrax arboreus*

Identification - An animal with the superficial appearance of a rodent that is thought, by some, to be more closely related to the Elephant. The body is round, the head is short and pointed with stiff grey whiskers, rounded ears and dark prominent eyes. The coat consists of dense, soft fur and varies in colour from pale grey-brown to dark brown. The belly is yellowish white as is a variable amount of the dorsal area. They have pale eyebrows and whitish edging to the ears and lips. The tail is very short, the legs are short and slender and the feet have specially adapted pads to aid them when climbing.

Distribution & Habits - An inhabitant of rain forests and woodlands over much of the region, up to an altitude of 4000 metres. They spend most of the daylight hours hidden away in tree holes, emerging under cover of darkness to feed. They are very agile climbers. They are very vocal and call frequently throughout the night as a means of maintaining their respective territories. The call commences with a series of groans and creeks, rising to a climax of eerie screams and shrieks. They feed on leaves, bark, shoots, fruits and grasses, as well as taking the occasional bird's egg and insect. Although the senses of sight, smell and hearing are acute they are often taken as prey by large owls and nocturnal cats.

KEY FACTS

Size	Height: 30 cms.
	Length: 60 cms.
	Weight: 3.5 kgs. (Average)
Breeding	Gestation: 8 months.
	Young: 1 or 2, rarely 3.
	Sexual Maturity: 16 months.
	Births: No specific season, although slight peaks occur during dry seasons.
Lifespan	Up to 12 years.
Lifestyle	Family: Solitary or pairs with young.
	Diet: Herbivorous.
	Main Predators: All cats, birds of prey and snakes.
	Habitat: Forests and woodlands of all kinds.

Conservation & Status - A common and widespread species throughout East Africa. In some areas they are hunted for their fur.

Rock Hyrax

Order	Family	Genus & Species
Hyracoidea	*Procaviidae*	*Heterohyrax brucei*

Identification - Easily mistaken at first glance for a rodent, this species is thought, by some, to be the closest living relative to the Elephant. Slightly smaller than the Tree Hyrax the body is round and the head short with a blunted snout and stiff grey whiskers. The ears are rounded and the legs and tail are short. The feet have specially adapted paws, the bottoms of which act like suction pads, aiding the animal as it runs among rocks and stones. The coat is dense and varies in colour from light to dark brown, often with the shoulders and legs lighter. They frequently have a patch of pale ochre-brown fur along the centre of the back. The underside is yellowish white. Like all hyrax species they have large incisor teeth.

Distribution & Habits - A species found in the vicinity of rocky outcrops and cliffs throughout East Africa. They live in medium to large colonies, seeking shelter and safety in the holes and cracks between the boulders. They can often be found in the early morning sunbathing and grooming. They feed on a variety of vegetation including leaves, bark, fruits, twigs, grasses and occasionally invertebrates. They are active throughout the day as well as on moonlit nights. The senses of sight, smell and hearing are acute and when danger threatens they issue a loud warning whistle which sends all members of the colony scurrying for safety among the rocks. They are the favourite prey item of the Verreaux's or Black Eagle.

KEY FACTS

Size	Height: 25 cms.
	Length: 57 cms.
	Weight: 3.2 kgs. (average).
Breeding	Gestation: 8 months.
	Young: 1 or 2, occasionally 3.
	Sexual Maturity: 16 months.
	Births: No specific season, but peak occurs from Feb-March
Lifespan	Up to 12 years.
Lifestyle	Family: Medium to large colonies.
	Diet: Herbivorous.
	Main Predators: All cats, large birds of prey and snakes.
	Habitat: Rocky outcrops and cliffs.

Conservation & Status - A common species throughout the region.

African Elephant

Order	Family	Genus & Species
Proboscidea	*Elephantidae*	*Loxodonta africana*

Identification - Quite unmistakeable, the worlds' largest land mammal. The elephant is so well known that a detailed description seems quite unnecessary, however, there are many interesting facts worth mentioning. The trunk is unique, being not only a nose, but also acting as an additional limb, with an extremely sensitive and flexible tip. An elephant will also use it's trunk for drinking, by sucking up large quantities of water before squirting it into the mouth. The large ears act as a very effective cooling system. The backs of the ears are covered by a network of blood vessels and the constant flapping helps the animal to regulate it's body temperature through evaporation. The tusks grow throughout the animals life, with the rate of growth varying enormously among individuals. An elephants eyesight is poor but the senses of smell and hearing are very acute. In spite of their massive bulk, soft cushioned undersides to their feet enable almost noiseless movement.

Distribution & Habits. Over recent years the number and distribution of the elephant has been greatly reduced, the main reasons being the expansion of human settlements and the poaching of ivory. They favour areas of wooded savannah and forests. Elephants eat a wide variety of vegetation from grasses and herbage to bushes and trees. In woodland areas considerable damage can be caused, even to mature trees, by the passage of a large herd. They are active throughout the day and night, although shelter is usually sought during the hottest part of the day. Most females and calves live together in herds averaging 10 to 20 animals, usually led by an old female they are very protective of their young. By contrast old bulls tend to lead a solitary existence.

KEY FACTS

Size	Height: 3.2 metres (average).
	Weight: Up to 6000 kgs.
Breeding	Gestation: 22 months.
	Young: One only.
	Sexual Maturity: 15 years.
	Births: No specific season.
Lifespan	Up to 70 years.
Lifestyle	Family: Herds of 10 to 20 on average, larger herds occasionally. Old bulls usually solitary.
	Diet: Herbivorous.
	Main Predators: Man. Unguarded calves may be taken by lions.
	Habitat: Savannahs and woodlands.

Conservation & Status - Under continual threat from poachers for their ivory. Their range and distribution has been much reduced in recent years due to this constant persecution. In some regions, as a result of vigourous anti-poaching patrols by the authorities, numbers are recovering.

Aardvark

Order	Family	Genus & Species
Tubulidentata	*Orycteropodidae*	*Orycteropus afer*

Identification - A rather bizarre, thick-set, short-legged animal that would be difficult to confuse with any other living mammal in East Africa. They are grey/brown in colour, with a sparse covering of bristly hair. Also known as the Ant Bear, they have an elongated muzzle, with a soft flexible snout and a long sticky tongue. The short legs are extremely powerful and have long claws that enable them to dig with ease. Their long pointed ears help to provide them with excellent hearing, they also have a well developed sense of smell, but rather poor eyesight. The tail is long, thick and powerful, resembling that of a Kangaroo and in the female, usually terminating in a white tip.

Distribution & Habits - Aardvarks are shy, solitary creatures, being rarely seen, as a result of their purely nocturnal habits. They spend the daylight hours sleeping in deep, self excavated burrows. These burrows may be single hole affairs or a more complicated arrangement of inter-twining tunnels and chambers. They can be found in areas of open savannah and light woodlands, with sandy or soft soils. They can range over large areas in search of food, covering distances up to ten kilometres in a night. They will visit dozens of termite mounds in turn, the long sticky tongue being used to collect ants, termites and other insects, which form the bulk of their diet. They will occasionally feed on soft, wild fruits. Single young, very rarely twins, are born in a burrow and will start to follow the female into the outside world at about two weeks old, they will remain with the female until the arrival of her next offspring.

KEY FACTS

Size	Length: 190 cms. (average)
	Weight: 67 kgs. (average)
Breeding	Gestation: 7 months.
	Young: One only, rarely twins.
	Sexual Maturity: About 2 years.
	Births: Some evidence to suggest a peak in the rainy season.
Lifespan	About 18 years.
Lifestyle	Family: Solitary - females with young.
	Diet: Omnivorous.
	Main Predators: Lions, Leopards, Spotted Hyaenas.
	Habitat: Open savannah & light woodlands, wherever termites are present.

Conservation & Status - Quite common in areas of high termite and insect densities. There is some evidence to suggest that the increase and spread of domestic livestock has provided new ranges for the Aardvark. The increased amount of dung created by livestock causing an increase in the invertebrate populations, which in turn provide food for the Aardvarks.

Unstriped Ground Squirrel

Order	Family	Genus & Species
Rodentia	*Sciuridae*	*Xerus rutilus*

Identification - This species is a true ground squirrel, never climbing trees. They have a thick coat of coarse hair. The head is brown with the face and inner edges of the forelegs pale tawny. The upper back is grey/brown. The coarse hair on the underside of the body is pale grey, almost white. The bushy tail, which is almost as long as the body, is a mixture of grey and black. The ears are small and the large dark eyes are surrounded by a ring of white fur. The feet have very long claws which are used for digging.

Distribution & Habits - A species of dry, semi-desert habitats throughout Kenya, Somalia and North-eastern Tanzania. They live in self excavated burrows, which can take the form of a single straight tunnel, or develop into a labyrinth of tunnels, chambers and entrance holes. They can be seen throughout the day scurrying about for food, often stopping and standing erect on hind legs to check for danger. They feed on a wide variety of vegetation including bulbs, roots and fruits. They occasionally come into conflict with farmers due to their taking maize, grain, yams and other cultivated crops. They also occasionally feed on insects. When feeding in the heat of the day, they often arc the tail above and over the back, this acts like a parasol and shades them from the fierce sun.

KEY FACTS

Size	Length including tail: 40 cms (average).
	Weight: 300 gms (average)
Breeding	Gestation: Unknown.
	Young: 2 - 6.
	Sexual Maturity: 6 months.
	Births: No specific season.
Lifespan	About 6 years.
Lifestyle	Family: Solitary.
	Diet: Roots, bulbs, leaves & insects.
	Main Predators: The smaller cats and
	large birds of prey.
	Habitat: Dry, semi-desert regions.

Conservation & Status - Reasonably common and in no immediate danger. Occasionally hunted by native peoples for food and by farmers to control damage to crops.

Striped Ground Squirrel

Order	Family	Genus & Species
Rodentia	*Sciuridae*	*Xerus erythropus*

Identification - Although larger than the Unstriped Ground Squirrel this species can at first glance look remarkably similar. The upper pelage colour varies from sandy to darkish brown, the bushy tail, which is almost as long as the body is grey/brown. The fur on the underside of the body is whitish grey. The main distinguishing feature, other than the size, is the presence of a white stripe which extends along the sides of the body from the rear of the forelimbs. The feet have long claws which aid the animal in digging.

Distribution & Habits - Usually found in wetter habitats than those occupied by the Unstriped Ground Squirrel, in open woodlands and rocky areas. They often move quickly into areas of forest recently cleared for farming, thereby coming into conflict with farmers by eating cultivated crops. They usually excavate their own burrows but will modify termite mounds and live happily among holes and crevices on rocky hill sides. There are generally several entrances to a burrow, the inner chamber being lined with dried grasses. They feed mainly on roots, bulbs, seeds and fallen fruits, although on occasions they may scramble up into fruiting bushes to plunder the fruit. They will also feed on birds eggs and young, as well as a variety of invertebrates. They move around in a casual, unhurried fashion, occasionally pausing to stand erect on their hind legs to note any approaching danger.

KEY FACTS

Size	Length including tail: 60 cms (average)
	Weight: 750 gms (average)
Breeding	Gestation: Unknown.
	Young: 2 - 6
	Sexual Maturity: 6 months.
	Births: No specific season.
Lifespan	About 6 years.
Lifestyle	Family: Generally solitary.
	Diet: Omnivorous.
	Main Predators: The small cats, snakes & Jackals.
	Habitat: Varied, open woodlands, rocky areas and the edges of cultivated farmland.

Conservation & Status - Quite common and in no immediate danger. They are occasionally controlled by farmers to lessen damage caused to cultivated crops.

Crested Porcupine

Order	Family	Genus & Species
Rodentia	*Hystricidae*	*Hystrix cristata*

Identification - The largest rodent in East Africa, a most distinctive animal making identification easy. The body coloration is blackish/brown and grey. The upper parts of the body, particularly the back and tail, are covered with long, sharp, black and white quills, which afford the animal good protection against attack from most predators. The belief that the quills can be 'shot' by the animal is a myth. A very stout looking creature with a large head and thick neck, from which grows a long, backward facing mane of bristles. The legs and tail, although substantial, are often hidden from view when the animal is in motion, by the spread and fall of the quills. They have long, sturdy claws which aid them greatly in digging burrows and in finding food.

Distribution & Habits - Widespread throughout the region, although their mainly nocturnal habits make them difficult to see. They can be encountered in a wide variety of habitats including forests, scrub and semi-desert regions, singly, in pairs or small groups. They spend the daylight hours sleeping in burrows, but on occasions may be seen sunbathing outside the burrow entrance. They do dig burrows for themselves but will also use holes excavated by Aardvarks as well as making good use of natural caves and crevices amongst piles of boulders. They have poor eyesight but well developed senses of hearing and smell. Their food consists mainly of vegetable matter including roots, tubers, barks and fruits. They will also gnaw bones, possibly as a source of calcium, but with the added benefit of sharpening their teeth. They will also occasionally take carrion. In some areas they are considered quite a pest, damaging and eating cultivated crops.

KEY FACTS

Size	Length including tail: 85 cms (average)
	Weight: 26 kgs
Breeding	Gestation: 2 months.
	Young: Usually 2 (1 - 4)
	Sexual Maturity: 2 years
	Births: Throughout the year.
Lifespan	Up to 20 years
Lifestyle	Family: Singly or small family parties.
	Diet: Omnivorous.
	Main Predators: Lions and Leopards very occasionally.
	Habitat: Forests, scrub and dry regions.

Conservation & Status - Widespread and quite numerous throughout much of the region. They are occasionally controlled by farmers protecting their crops.

African Hare

Order	Family	Genus & Species
Lagomorpha	*Leporidae*	*Lepus capensis*

Identification - A common animal with a coat of grizzled brown fur. The legs are long and slender, the hind legs, being nearly twice the length of the forelegs, allowing the animal to run and jump at high speed. The legs are covered with pale rufous fur while the fur on the underside of the body is whitish/grey. The ears are very long, with dark edges and tipped with black. They have long white whiskers and large, prominent golden-brown eyes. The tail is short and fluffy, white below and black above.

Distribution & Habits - Widely distributed in open, dry savannah habitats. They usually keep themselves well hidden amongst vegetation during the daytime, becoming more active in the late afternoon, the evening and throughout the night. They have good eyesight and the senses of hearing and smell are very well developed. They feed on a wide variety of vegetation including leaves, twigs, barks, berries, fruits, roots, tubers and grasses. They will often sit upright on their hind legs, particularly in long grass, on the lookout for danger. When any danger does approach they rely heavily on camouflage as a means of avoiding initial detection. If, however, a predator wanders too close, the hare will burst from cover at high speed, often exceeding 70 kph. If pursued they will often seek refuge in the tunnels of burrowing animals such as Aardvark.

KEY FACTS

Size	Length including tail: 55 cms (average)
	Weight: 1.75 kgs
Breeding	Gestation: 1.5 months.
	Young: 1 - 6 Usually at least 2 litters a year.
	Sexual Maturity: About 8 months
	Births: No specific season
Lifespan	Up to 12 years
Lifestyle	Family: Solitary
	Diet: Herbivorous.
	Main Predators: Numerous, including Cheetah, Jackal, Owls and large birds of prey.
	Habitat: Open, dry savannah regions.

Conservation & Status - Very common in suitable habitats.

Side Striped Jackal

Order	Family	Genus & Species
Carnivora	*Canidae*	*Canis adustus*

Identification - The least common of the three species of jackal to be found in East Africa. It bears the features typical of the family, sleek body, long legs and a bushy tail. The drab coat colour is a mixture of rufous, grey and black. The tail is blackish with a pronounced white tip, which is not present in the other two species, making it a useful field characteristic. The side stripes vary in prominence from animal to animal, but are usually clearly discernable as a whitish line edged with black extending from the shoulder, along the body towards the base of the tail. The muzzle is blunter than in other jackal species and the ears are smaller.

Distribution & Habits - Found in a variety of habitats including woodland, bush, scrub and open grasslands. They have a varied diet which includes rodents, small birds and mammals, invertebrates ranging from large beetles and crickets to termites, as well as carrion and wild fruits. They can be seen during daylight hours but usually rest up in dense cover, becoming more active during the night and just prior to sunrise. Side Striped Jackals pair for life, with both male and female sharing in the rearing of offspring. Although they will dig tunnel dens for themselves, they will readily take over old Aardvark holes, holes in termite mounds and cavities among hill side boulders. The senses of sight, hearing and smell are all very well developed.

KEY FACTS

Size	Length including tail: 120 cms.
	Height at shoulder: 50 cms.
	Weight: 12 kgs (average)
Breeding	Gestation: Around 2 months
	Young: 1 - 4
	Sexual Maturity: Unknown
	Births: At any time of year, but there is some evidence of a peak at the onset of the rainy season.
Lifespan	Up to 12 years
Lifestyle	Family: Pairs and small family parties.
	Diet: Omnivorous.
	Main Predators: Large cats.
	Habitat: Woodlands, scrub and open grasslands.

Conservation & Status - Quite common and not under immediate threat. Many are killed by speeding vehicles on roads and tracks.

Black-Backed Jackal

Order	Family	Genus & Species
Carnivora	*Canidae*	*Canis mesomelas*

Identification - The commonest and most handsome of the three species of jackal found in East Africa. The body is slender, the legs are long and the head is fox-like with large pointed ears. The head, the lower portion of the body and the legs are pale rufous in colour. Along the hindneck and back runs a saddle of black fur, flecked with a variable amount of white. The tail is bushy, pale rufous and white edged and tipped with black. The underparts are whitish. Young animals are less well marked than adults, being for the most part grey/brown.

Distribution & Habits - Found throughout the region in open savannah, light woodland and areas of scrub and bush. They are active during both the day and the night, but usually seek shade and seclusion during the heat of the day. Although they scavenge from kills made by the large carnivores their main source of food is derived from small birds and mammals, lizards and a variety of insects and fruits. They will also hunt small antelopes, often doing so in small packs. They will create dens in old Aardvark burrows or in dense vegetation. Like the Side-striped Jackal they pair for life and share the burdens of raising their offspring. The young will remain with their parents for about 8 months after which, with parental encouragement, they will move on to establish a territory for themselves. In some districts they are becoming something of a pest, often stealing lambs and poultry from in and around settlements. The senses of sight, hearing and smell are very acute.

KEY FACTS

Size	Length including tail: 110 cms.
	Height at shoulder: 45 cms.
	Weight: 10 kgs (average)
Breeding	Gestation: 2 months.
	Young: 5 - 6 average (2 - 10)
	Sexual Maturity: 8 months.
	Births: Peak occurs after the rains.
Lifespan	Up to 12 years.
Lifestyle	Family: Pairs or family parties.
	Diet: Omnivorous.
	Main Predators: Large cats, pythons.
	Habitat: Open savannah and light woodlands.

Conservation & Status - Common throughout most of the region.

Common Jackal

Order	Family	Genus & Species
Carnivora	*Canidae*	*Canis aureus*

Identification - The Common Jackal shares the same fox-like features as the Black-backed Jackal, but has a heavier coat which is subject to some variation in colour. Usually the upper side is uniform dull yellow/grey. The chest and underparts are whitish, with the legs and the backs of the ears showing some rufous/yellow. The bushy tail is yellow/grey tipped with black. Young animals appear grey/brown and rather drab. There have been sightings of animals with blackish coats.

Distribution & Habits - Found in northern Tanzania, north and eastern Kenya and Somalia, favouring areas of open grassland, rocky hill sides and patches of bush and scrub. A species that can readily be seen during the day, but is more active under cover of darkness. The senses of sight, hearing and smell are very well developed. Like other jackal species they feed on a wide variety of items including, small birds and mammals, particularly rodents, insects and some fruits as well as taking carrion. They will occasionally attempt to hunt small antelope, being far more successful when joining forces with other individuals. They will often be found scavenging around settlements and in cities during the night, taking any form of discarded food or garbage. They appear to scavenge less than other jackal species from kills of larger predators, often being chased away by Black-backs and Hyaenas. They will, however, feed on the afterbirths present in vast quantities during the wildebeest calving season in February.

KEY FACTS

Size	Length including tail: 110 cms (average)
	Height at shoulder: 45 cms (average)
	Weight: 11 kgs (average)
Breeding	Gestation: 2 months.
	Young: Variable litter sizes from 3 - 8.
	Sexual Maturity: About 10 months.
	Births: No specific season.
Lifespan	Up to 12 years.
Lifestyle	Family: Singly, pairs or family parties.
	Diet: Omnivorous.
	Main Predators: Large cats. Large birds of prey will take cubs.
	Habitat: Open grasslands and areas of bush and scrub.

Conservation & Status - Common over much of the region, particularly on the plains of the Serengeti.

Hunting Dog

Order	Family	Genus & Species
Carnivora	*Canidae*	*Lycaon pictus*

Identification - The coat pattern and coloration of hunting dogs shows great individual variation, but is usually that of dark brown/black with irregular blotches and patches of tan, cream and white. They have a long slender body, long legs, large, rounded ears and a bushy, white tipped tail. The head is rather short and broad, with powerful jaws. Young pups are blackish/brown with individual and irregular white patches, most prominent on the legs. The tail is tipped white.

Distribution & Habits - Once common on the open grassland plains of East Africa, their distribution has been drastically reduced in recent years as a result of disease and persecution by man for stock stealing. Hunting as a pack they are extremely successful. They pursue their chosen victim relentlessly, until as a result of fatigue and by sheer weight of numbers they pull it to the ground. They feed mainly on antelope species but are capable of taking prey as large as zebra. They hunt mainly at dusk and dawn, resting up during the heat of the day. They are a nomadic species covering vast distances during their persistent wanderings over territories up to 2000 sq, kms in extent. They only remain in the same area when the alpha female gives birth to a litter of pups. The chosen den is usually an old Aardvark or Warthog burrow and large litters of up to 18 pups have been recorded. Once the pups are weaned, the entire pack will help with their feeding and raising. At about 10 weeks old, the pups, along with the pack, will leave the den area and again take up a nomadic existence.

KEY FACTS

Size	Length including tail: 150 cms
	Height at shoulder: 75 cms
	Weight: 27 kgs (average)
Breeding	Gestation: 2.5 months.
	Young: 10 (average)(2-18)
	Sexual Maturity: 18 months.
	Births: No specific season. Slight peak at the onset of the rainy season.
Lifespan	10+ years
Lifestyle	Family: Packs of 6-20
	Diet: Carnivorous.
	Main Predator: Man
	Habitat: Open grassland plains and bush.

Conservation & Status - The drastic decline in numbers over recent years, as a result of canine diseases such as distemper and rabies, by persecution and destruction of habitat, has brought this magnificent animal to the verge of extinction in the region.

Bat-Eared Fox

Order	Family	Genus & Species
Carnivora	*Canidae*	*Otocyon megalotis*

Identification - This delightful little animal has a slightly shaggy, uniform coat of greyish/yellow, paler on the flanks and underside. The lower portion of the legs, the tips of the enormous rounded ears, the nasal area of the muzzle and the tip of the large bushy tail are black. The eyes are set in patches of black fur which are joined across the bridge of the muzzle, giving the animal the appearance of wearing a 'highwayman's mask'. The body is rounded at the rear and higher in the hindquarters than at the shoulders.

Distribution & Habits - An inhabitant of open savannahs, sandy areas and light woodlands. They often dig their own burrows, but will readily occupy tunnels and chambers vacated by other small, burrowing mammals. They are active mainly at dawn, dusk and throughout the night, but will often spend a considerable time lying around their den entrances during the hours of daylight. They feed mainly on small mammals, lizards, snakes and a multitude of insects including beetles, scorpions and termites. They will also take wild fruits and berries in season, as well as digging for roots and tubers. They mate for life and live in small family parties. Litter sizes vary from 2 to 6 and the development of the pups is rapid. By 4 weeks they are taking solid food in the form of insects. Whilst feeding they are continually on the lookout for danger, which usually approaches from above in the shape of large eagles.

KEY FACTS

Size	Length including tail: 90 cms
	Height at shoulder: 30 cms
	Weight: 5 kgs (average)
Breeding	Gestation: 2 months
	Young: Usually 3 or 4 (2-6)
	Sexual Maturity: 9 months
	Births: Most seem to occur between Oct & March.
Lifespan	Up to 13 years
Lifestyle	Family: Pairs or family parties.
	Diet: Omnivorous.
	Main Predators: Large birds of prey, Hyaenas.
	Habitat: Open savannah and light woodlands.

Conservation & Status - Quite widespread and common. However, in some areas numbers are declining as a result of the spread of human settlements. They are also hunted by some native peoples as well as being susceptible to many canine diseases which, from time to time, causes drastic reductions in their number.

Honey Badger

Order	Family	Genus & Species
Carnivora	*Mustelidae*	*Mellivora capensis*

Identification - A robust, stocky animal with thick, loose skin, a long body and short, sturdy legs. The honey badger moves with a purposeful lumbering motion. The coat consists of short, coarse hair, the colour of which is evenly divided into two sections. The crown of the head and the upper portion of the body and tail are greyish/white, while the remaining areas are jet black. Young animals are brownish. The head is large with very powerful jaws and small ears. The front paws are equipped with large, robust claws.

Distribution & Habits - Found throughout the region over a wide range of habitats from open grasslands to forests. They are mainly nocturnal but are sometimes encountered at dawn and dusk. They live in burrows which are easily excavated with the large front claws, they will also readily occupy holes originally dug by Aardvarks. They feed on a wide variety of foods including small mammals, snakes, invertebrates, roots, tubers, various wild fruits and the pupae and honey of wild bees. A strange association has developed between the honey badger and a small bird called a honeyguide. On locating a bees nest the honeyguide, by the use of short display flights and calls, will lead a badger to the nest. The badger easily rips open the nest and feasts on the contents, after which the honeyguide too takes it's fill. The honey badger is a tenacious and fearless creature and will attack animals many times it's own size when threatened. They are generally thought of as being shy, solitary animals, although very little is known about much of their social life.

KEY FACTS

Size	Length including tail: 100 cms
	Weight: 11 kgs (average).
Breeding	Gestation: 6-7 months.
	Young: 1 or 2
	Sexual Maturity: Unknown.
	Births: No specific season.
Lifespan	Up to 24 years.
Lifestyle	Family: Singly or in pairs.
	Diet: Omnivorous.
	Main Predators: Very few, other than man.
	Habitat: Found in most habitats.

Conservation & Status - Although widespread they are by no means common. They are often trapped or snared by local people, in an attempt to minimise the raiding of their bee hives.

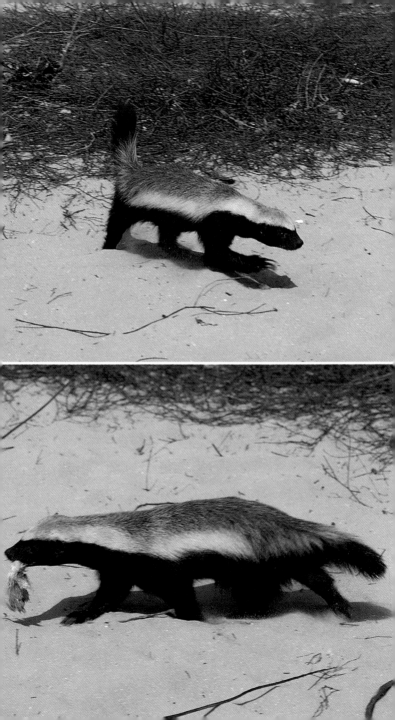

Large-Spotted Genet

Order	Family	Genus & Species
Carnivora	*Viverridae*	*Genetta tigrina*

Identification - A very lithe and beautiful small carnivore. The body is long and sleek, the legs are short and the paws have sharp, retractile claws. The short coat is greyish/yellow boldly marked with a black dorsal stripe, either side of which extend several rows of large black spots. These spots are occasionally edged with rufous. The short, pointed face has white patches on the upper cheeks and around the nasal area, large, honey-coloured eyes and pointed ears, the insides of which often show as naked pink flesh. The sides of the muzzle are black. The bushy tail is almost as long as the body and is ringed alternately with grey and black bands, terminating in a black tip. Young animals are generally greyer than adults with fainter markings.

Distribution & Habits - Found throughout the region, favouring areas of woodlands, forests and thickets, usually in the vicinity of water. They often frequent safari lodges as well as other human settlements, on occasions causing considerable damage to poultry stocks. Their diet is extremely varied and includes rodents, lizards, snakes, frogs, large insects such as grasshoppers and crickets, plus a variety of wild fruits. They are nocturnal creatures, spending the daylight hours in burrows or hollow trees. They are very agile and accomplished climbers, but are equally suited to life on the ground where they seek much of their prey. The senses of sight, smell and hearing are well developed. Litters of kittens usually number 2 or 3 and by 3 months they are hunting for themselves, by 6 months they are completely independent of their parents.

KEY FACTS

Size	Length including tail: 100 cms
	Weight: 3 Kgs
Breeding	Gestation: 2.5 months.
	Young: 1-4
	Sexual Maturity: About 2 years.
	Births: No specific season.
Lifespan	Up to 9 years.
Lifestyle	Family: Singly or in pairs.
	Diet: Omnivorous.
	Main Predators: Little is known, but the authors would suspect that the young in particular are susceptible to attack by other nocturnal, predatory mammals and by large owls.
	Habitat: Woodland and forest areas.

Conservation & Status - Widespread throughout the region, the population would appear to be stable and under no threat, other than from the destruction of habitat.

African Civet

Order	Family	Genus & Species
Carnivora	*Viverridae*	*Civettictis civetta*

Identification - A short legged animal of sturdy build with a long bushy tail, which is ringed with alternate bands of grey and black and has a black tip. The general coat colour is grizzled grey, marked with a varied selection of spots and stripes. In motion the head is carried low and the shoulders are lower than the hindquarters. The coat is shaggy and loose, with the semblance of a mane extending from the neck to the root of the tail. A black dorsal stripe extends from the crown of the head to the tail, the throat, chest and lower portion of the face and limbs are black. Two bands of black fur extend from the ears, down the sides of the neck, to the chest. The whole body is covered with a series of irregular black spots, the lower part of the hind limbs sometimes showing horizontal black stripes. Young animals are browner with only faint markings.

Distribution & Habits - Found throughout much of the region over a variety of habitats including open savannahs, woodlands and forests, they avoid desert areas. A nocturnal animal spending the daylight hours sleeping in a burrow, in dense grasses and thickets or other suitable hiding places. Their food is varied and includes small to medium sized mammals, reptiles, amphibians, berries, fruits and carrion. On occasions they will inflict serious damage to poultry stocks and are often hunted and trapped by man as a result. Eyesight is relatively poor, whilst hearing is good and the sense of smell very acute. They scent mark their surroundings with musk secreted from a gland beneath the tail. For many centuries civets were kept in captivity and 'milked' of their musk, which was found to be an effective fixative in the production of flower based perfumes. In more recent years chemical substitutes have been found.

KEY FACTS

Size	Length including tail: 130 cms
	Weight: 12 kgs (average)
Breeding	Gestation: 2.5 months.
	Young: 1 - 4
	Sexual Maturity: 12 months.
	Births: Peak births occur at the onset of the rainy season.
Lifespan	Up to 14 years.
Lifestyle	Family: Solitary, occasionally in pairs or family parties.
	Diet: Omnivorous.
	Main Predators: Man.
	Habitat: Open savannahs, woodlands and forests.

Conservation & Status - Quite widespread and under no immediate threat. In some regions they are hunted by local people, with the aid of dogs and nets, in retaliation for poultry raiding.

Slender Mongoose

Order	Family	Genus & Species
Carnivora	*Viverridae*	*Herpestes sanguineus*

Identification - An animal with a long, slender body, short legs, long pointed face and a black tipped tail almost as long as the body. The coat is formed of fine hair varying in colour from light grey to darkish brown, often paler on the underside. The coat colour would appear to vary as a result of the type of habitat occupied, being darker in forest and mountain areas and lighter in drier more open locations.

Distribution & Habits - Widespread throughout the region in a variety of habitats including lightly wooded savannah, woodland, dense mountain forests, to an altitude of 2500 - 3000 metres, rocky hill sides and cultivated farmland and settlements. They can be seen throughout the day. They feed on small birds and mammals, lizards, snakes, invertebrates and occasionally wild fruits. Although they are able to climb trees quite well, they spend the vast majority of their time foraging on the ground. They occasionally cause damage to poultry stocks. They move with a lithe, rippling motion, holding the tail high, often stopping to check for approaching danger, usually in the form of large birds of prey, by rearing up on their hind legs. The senses of sight, hearing and smell are very well developed.

KEY FACTS

Size	Length including tail: 60 cms (average)
	Weight: 500 gms (average) Males are heavier than females by as much as 20%.
Breeding	Gestation: About 2 months.
	Young: 2 - 4
	Sexual Maturity: 12 months.
	Births: Not known.
Lifespan	Not known.
Lifestyle	Family: Singly or in pairs.
	Diet: Omnivorous.
	Main Predators: Large bird of prey and snakes.
	Habitat: Wooded savannah, woodlands and forests.

Conservation & Status - A widespread and successful mongoose species, which appears to be under little threat other than from destruction of habitat.

Banded Mongoose

Order	Family	Genus & Species
Cornivora	*Viverridae*	*Mungos mungo*

Identification - A well built, sturdy mongoose with a broad head and pointed face typical of the family. The hind quarters are rounded and higher than the forequarters. The tail is quite long, about half the length of the body, and tapers to a pointed tip. The coarse coat is grey/brown with a series of dark brown, almost black vertical stripes on the back and hindquarters. The lower portion of the legs and the tip of the tail are black. The ears are small. The feet are equipped with long claws which aid the animal greatly in searching out subterranean food items.

Distribution & Habits - A highly social animal, found throughout the range in areas of open savannah, often in the vicinity of water. They are active throughout the day, usually foraging in large family parties across the plains, in search of food which consists of a wide variety of invertebrates, lizards, snakes, small rodents and amphibians, as well as occasional roots and fruits. Whilst foraging they are constantly on the look out for danger often raising themselves into an upright position on their hind legs to gain a better view. An alarm call from one member of the pack will have the whole group scurrying for cover. They are territorial animals and are extremely hostile towards neighbouring packs that dare to trespass. They have many den sites throughout a territory, usually moving to a new location every few weeks. Their preferred den sites are situated in large termite mounds. The senses of smell, hearing and sight are well developed.

KEY FACTS

Size	Length including tail: 60 cms (average).
	Weight: 2 Kgs (average).
Breeding	Gestation: 2 months.
	Young: 2 - 6
	Sexual Maturity: 12 months.
	Births: Peak occurs during the rainy season.
Lifespan	Up to 10 years.
Lifestyle	Family: Parties of up to 40 animals.
	Diet: Omnivorous.
	Main Predators: Cats and large birds of prey.
	Habitat: Open savannahs.

Conservation & Status - Common and widespread in suitable habitats.

Dwarf Mongoose

Order	Family	Genus & Species
Carnivora	*Viverridae*	*Helogale undulata*

Identification - As it's name implies, this is the smallest mongoose species found in East Africa. It retains all the main features typical of the group, a broad head and pointed face, small ears, short legs and long claws. The tail is about half the length of the body and tapers to a point. The fine coat can be variable in colour, from reddish brown to grizzled grey, often lighter below and with darker legs and feet. They utter a wide range of whistles, squeaks and other noises which keep the pack in contact at all times.

Distribution & Habits - Found over much of the region, favouring dry grasslands, woodlands and areas of scrub and bush. They are often common around safari lodges and dwellings, where they become quite tame. They live together in family parties usually numbering between 6 - 15 individuals, occupying termite mounds, natural crevices between boulders, hollow trees and other suitable safe denning locations. They are active throughout the day, foraging as a group but fanning out and feeding individually. They are continually on the look out for danger, an alarm call from any individual instantly alerting the whole pack. They feed mainly on insects, lizards, small birds and mammals and wild fruits. They can survive without water, obtaining all of their moisture requirement from their food but, if water is available they will make use of it. Each pack is led by a single breeding pair, the female being dominant, with all the other pack members helping to raise the offspring. They are very territorial and will defend their ground fiercely against other intruding packs.

KEY FACTS

Size	Length including tail: 40 cms (average)
	Weight: 300 gms (average)
Breeding	Gestation: About 50 days.
	Young: 2 - 6
	Sexual Maturity: 12 months.
	Births: Peak occurs during the rainy seasons.
Lifespan	Up to 10 years.
Lifestyle	Family: Family parties usually 6 -15 strong.
	Diet: Omnivorous.
	Main Predators: Birds of prey and snakes.
	Habitat: Open grassland, bush and scrub and woodlands.

Conservation & Status - Widespread and quite common. They are hunted by local people in some regions for food and by farmers in retaliation for poultry raiding.

Aardwolf

Order	Family	Genus & Species
Carnivora	*Hyaenidae*	*Proteles cristatus*

Identification - The Aardwolf has the appearance of a small Hyaena. The most prominent identification feature is the long, thick mane of hair which extends along the entire length of the back, from the rear of the ears to the tail. The head has a long muzzle and large pointed ears. The legs are long and slender. The tail is long and bushy. The colour of the coat is yellowish brown with yellow/white throat, cheeks and underparts. The body has an irregular pattern of broken vertical black stripes, the stripes on the forelimbs are horizontal. Like the Hyaenas, the Aardwolf has a sloping back, being higher at the shoulders than at the hindquarters.

Distribution & Habits - Found throughout much of Kenya, Tanzania and northern Uganda, favouring areas of open dry plains and bush. They avoid mountainous areas. Their main distribution is, to a great extent, linked to that of Harvester Termites. These termites prosper in areas that are heavily grazed by plains game animals and domestic stock, as a result the increase in human population and consequently that of domestic animals may well benefit the Aardwolf. They are extremely secretive animals, foraging at dusk and throughout the night, during the daytime they usually remain hidden in underground burrows. They do excavate burrows for themselves, but will readily occupy holes vacated by other burrowing creatures such as Aardvark. Their main food intake consists of termites, but they will take other insects, birds eggs and some small mammals and reptiles. They have good eyesight and a well developed sense of smell. Their hearing is very acute and much of their food is initially located by sound.

KEY FACTS

Size	Length: 95 cms including tail (average).
	Height: 50 cms.
	Weight: 12 kgs (average).
Breeding	Gestation: 3 months.
	Young: 2 - 4
	Sexual Maturity: Not known.
	Births: No specific season known.
Lifespan	Up to 13 years.
Lifestyle	Family: Solitary except when raising young.
	Diet: Carnivorous.
	Main Predators: Lions, Leopards, Hyaenas and large Pythons.
	Habitat: Dry regions, savannahs and bush.

Conservation & Status - Numbers are thought to be stable.

Striped Hyaena

Order	Family	Genus & Species
Carnivora	*Hyaenidae*	*Hyaena hyaena*

Identification - Smaller and less powerful than the Spotted Hyaena, the Striped Hyaena has many similar features, a large broad head, long legs, a sloping body from high shoulders to lower hind quarters and the familiar gait. The coat has a shaggy appearance with a long, buff/grey dorsal mane extending from the nape to the tail. The overall body colour is yellowish-grey, with vertical black stripes on the body, horizontal black stripes on the legs and a black throat and muzzle. The tail is long and bushy, appearing as a continuation of the dorsal mane, with a black tip. The ears are large and pointed and juvenile animals lack the shaggy mane but have a black dorsal stripe.

Distribution & Habits - Found in woodland, dry savannah and semi-desert areas from northern Tanzania to the north African coast. They are almost entirely nocturnal, remaining hidden during daylight hours in dense vegetation, in caves and rock crevices, or in subterranean burrows. They can walk considerable distances during a nights foraging, (average 20 kms), taking carrion, small mammals, invertebrates, lizards and wild fruits. They will also occasionally plunder cultivated crops. When a glut of food is located, surplus is often hidden among thick vegetation for consumption at a later date. The senses of smell, hearing and sight are very acute. Although they will drink regularly when water is available, they are able to survive for long periods of drought, obtaining sufficient moisture from their food.

KEY FACTS

Size	Length including tail: 140 cms (average).
	Height at shoulder: 75 cms.
	Weight: 45 kgs (average).
Breeding:	Gestation: 3 months.
	Young: Usually 2-4 (1-6)
	Sexual Maturity: 2 to 3 years.
	Births: No specific season.
Lifespan	Up to 20 years.
Lifestyle	Family: Singly or in pairs.
	Diet: Omnivorous.
	Main Predators: No common enemies. Keeps well clear of Lions and Spotted Hyaenas, it has a very active and pungent anal gland which other animals find offensive.
	Habitat: Dry woodland and savannah.

Conservation & Status - By no means common. Under threat from the spread of human population.

Spotted Hyaena

Order	Family	Genus & Species
Carnivora	*Hyaenidae*	*Crocuta crocuta*

Identification - The commonest large predator in East Africa. A very substantial animal, with well developed forequarters, a large broad head with rounded ears and extremely powerful jaws. The hindquarters are substantially lower than the forequarters resulting in a rather ungainly gait. The coat is short, with the semblance of a mane covering the lower neck and shoulders, grey/yellow in colour heavily marked with irregular black spots. The underside and chest are lighter and the lower portion of the legs are black. The coat usually has an unkempt appearance. The tail is of medium length with a black tip. The muzzle is dark brown/black.

Distribution & Habits - Common throughout the region in a variety of habitats, with the densest populations occurring on open savannahs where they live in large communities, some 'clans' numbering up to 80 individuals. Females are larger than males and dominate within the clans. They are extremely successful predators and, hunting as a pack, are capable of bringing down large herbivores. They will also regularly scavenge from the kills of lions and other cats. They are active mainly at night, when their eerie, wailing calls can be heard. They can very often be seen during the daytime resting outside their burrows, in thickets, long grass or in muddy waterholes. It is no easy task to determine the sex of individual animals, as the females have external genitals similar to those of the males. The senses of sight, smell and hearing are acute.

KEY FACTS

Size	Length including tail: 160 cms (average).
	Height at shoulder: 80 cms (average)
	Weight: 60-80 kgs, females are on average 10% heavier than males.
Breeding	Gestation: About 4 months.
	Young: 1 -3.
	Sexual Maturity: 2 to 3 years.
	Births: No specific season.
Lifespan	Up to 40 years.
Lifestyle	Family: Mainly in large packs,(clans).
	Diet: Omnivorous.
	Main Predators: Lions, Hunting Dogs & Man.
	Habitat: Mainly open savannah.

Conservation & Status - Still common and widespread in suitable habitats. During recent times numbers have declined, in some areas, as a result of human persecution, usually by poisoning.

Lion

Order	Family	Genus & Species
Carnivora	*Felidae*	*Panthera leo*

Identification - The largest and most powerful of East Africa's cats. The head is very broad with a short muzzle and small, rounded ears, the backs of which are black. The coat, with the exception of the mane, is very short, the ground colour being a sandy yellow often with faint spots, which are particularly noticeable in younger animals. The mane, which is only present on the males, is variable in colour from pale buff to black and covers the neck and shoulders. A full mane may take up to 6 years to develop.

Distribution & Habits - An animal of open savannahs, grassy plains and lightly wooded areas. They are unusual among cats in being very social, living in prides of 30 or more individuals. Lions are inactive for much of the day, often resting for up to 20 hours in the shade of trees and bushes. Their sight and smell are good and their hearing is exceptionally keen. Lions will hunt singly, in pairs or in large prides and, as a rule, the more lions taking part the larger the prey species. The hunting is undertaken primarily by the lioness's who secure around 80% of the prides food requirement. Lions will also scavenge whenever the opportunity arises. Although capable of running at speeds of up to 60 kph, lions have little stamina and give up most chases if they are not successful within 200 metres. On the plains of the Serengeti in Tanzania, many of the prides seem less territorial than in other areas, often following herds of plains animals on their annual migrations.

KEY FACTS

Size	Length including tail: 300 cms, males larger than females.
	Height at shoulder: 115 cms average, males larger than females.
	Weight: Males 190 kgs, females 150 kgs. (average).
Breeding	Gestation: 3.5 months.
	Young: Usually 2 or 3
	Sexual Maturity: 1.5 years.
	Births: No specific season.
Lifespan	15+ years.
Lifestyle	Family: Prides of 30 or more - small bachelor groups.
	Diet: Carnivorous.
	Main Predators: The young are vulnerable to Hyaenas and Hunting Dogs.
	Habitat: Open savannah and light woodlands.

Conservation & Status - Still common and widespread, although a decrease in numbers has occurred in recent years, due to the increase in human population and the subsequent spread of settlements reducing the available habitat.

Leopard

Order	Family	Genus & Species
Carnivora	*Felidae*	*Panthera pardus*

Identification - A muscular, thickset cat with short, powerful limbs. The head is broad with a muzzle of medium length, strong jaws and long, white whiskers. The ears are small and round, the backs of which are black with a prominent white marking in the centre. The coat is yellowish/tan and covered with black/brown spots, grouped in rosettes on the body but generally solid black on the head and lower legs. In some individuals the close grouping of the spots and rosettes gives an impression of a much darker pelage. The chin, throat and underside are off-white. There can be a great deal of regional variation in coat colour, length and density, in response to differing altitude and climatic conditions. The tail is long, spotted from the root to the centre but terminating with a series of black rings. The female is smaller and lighter than the male.

Distribution & Habits - Although not common the leopard is found over much of East Africa. Being mainly nocturnal they are seldom seen by day, but can be found wherever there is cover and sufficient food, preferring wooded savannahs and rocky outcrops. They are most likely to be encountered during the day resting in deep vegetation or, on occasions, in a tree. Hunting is mainly undertaken at dusk or during the night. They have remarkable vision, exceptional hearing and a good sense of smell. They prey on many mammal species particularly antelopes, they also have a particular liking for baboons. They have tremendous strength and, having made a kill, will often carry it high into a tree to evade the attention of other predators. Adult leopards lead a solitary existence, only coming together during periods of mating. Females will give birth to as many as six cubs which she keeps well hidden for the first six weeks. They will become independent at about 2 years old.

KEY FACTS

Size	Length including tail: 290 cms.
	Height: 60 cms (average).
	Weight: Males 63 kgs, females 45 kgs, (average).
Breeding	Gestation: 3.5 months.
	Young: 1-6
	Sexual Maturity: 2.5 years
	Births: No specific season.
Lifespan	About 20 years.
Lifestyle	Family: Solitary.
	Diet: Carnivorous.
	Main Predators: Lions and man.
	Habitat: Forest, bush and rocky outcrops.

Conservation & Status - By no means a common animal. Under pressure from man, often being poached for their skins.

Cheetah

Order	Family	Genus & Species
Carnivora	*Felidae*	*Acinonyx jubatus*

Identification - The world's fastest land mammal, built for sheer speed. The body is sleek and flexible, with a powerful chest and very long legs. The coat is yellow/buff, randomly marked with small black spots, the chest and underside are almost white. The small, rounded head has small ears, large, deep orange eyes, black spotting on the forehead and cheeks and prominent black stripes extending from the inner corners of the eyes to the mouth. The tail is long, yellow/buff spotted with black for much of it's length, but terminating with a series of black rings and a white tip. Each individual animal has a pattern of tail rings as unique as human finger prints. A mane of coarse fur is often discernable on the lower neck and shoulders, particularly in young animals.

Distribution & Habits - Found on dry, open savannahs and in areas of bush and scrub, where they hunt by sight throughout the day. They will often seek an elevated position, on a fallen tree, termite mound or rocky outcrop, when looking for prey. They will stalk as close as possible before attacking with a final burst of speed of up to 100 kph. This burst of speed can only be maintained for about 500 metres. Whilst running at full speed the cheetah will trip it's prey and seize it quickly by the throat in a suffocating grip. Having secured a kill they will attempt to drag the carcase into cover to avoid the attention of scavenging hyaenas and lions. They prey mainly on small to medium sized antelope, particularly Thomson's Gazelles, but will readily take hares and savannah birds. Cheetahs can live a solitary existence as well as in pairs or small family parties.

KEY FACTS

Size	Length including tail: 200 cms (average).
	Height at shoulder: 76 cms (average).
	Weight: 50 kgs (average). Males weigh about 20% more than females.
Breeding	Gestation: 3 months.
	Young: Usually 2-4 (1 - 6)
	Sexual Maturity: Around 1 to 1.5 years.
	Births: No specific season.
Lifespan	Around 15 years.
Lifestyle	Family: Singly, pairs or family parties.
	Diet:Carnivorous.
	Main Predators: Lions, Leopards and Spotted Hyaenas.
	Habitat: Open savannahs and dry bush.

Conservation and Status - Numbers have declined over recent years as a result of the increased human population and the consequent spread of settlements, resulting in a reduction of large, undisturbed areas suitable for cheetahs.

Caracal

Order	Family	Genus & Species
Carnivora	*Felidae*	*Felis caracal*

Identification - A tall, slender cat with a broad, flat head, large eyes and pointed, triangular ears with long tufts of black fur flowing from the tips. The backs of the ears are black, the insides are white edged with black. The coat colour is usually tawny/red, but on occasions dark almost black individuals occur. The chest, underside and insides of the legs are white, with some faint spotting usually discernable on the lower portion of the legs. The tail is relatively short. Males are usually larger than females. They are extremely fast and agile with powerful, well developed hindquarters which slope down to the shoulders.

Distribution & Habits - A shy rather retiring animal, rarely seen during the daytime, being almost entirely nocturnal. They inhabit areas of savannah, usually with kopjes or other natural features offering cover in which to establish dens. They will occupy old Aardvark burrows, caves and hollow trees. They are known to feed on a variety of mammals including small antelope, hares, rats and Rock Hyrax as well as birds which they often take during the night as the birds sleep or in flight when flushed from ground cover. They can cause considerable damage to goat, sheep and poultry stocks often coming into conflict with local farmers as a result. Their senses of sight and hearing are very keen, while their sense of smell would appear to be only moderate. They lead a solitary existence, holding territories covering several square miles.

KEY FACTS

Size	Length including tail: 100 cms (average).
	Height at shoulder: 50 cms.
	Weight: 17 kgs (average).
Breeding	Gestation: 70 days.
	Young: Usually 2 or 3
	Sexual Maturity: 2 years.
	Births: No specific season.
Lifespan	Around 16 years.
Lifestyle	Family: Solitary.
	Diet: Carnivorous.
	Main Predators: No solid evidence. Large predators may be a danger.
	Habitat: Open savannahs & semi-deserts.

Conservation & Status - Numbers seem to be stable, but there is still the need for more research to be undertaken, as much is to be learned about the social life of this shy cat.

Serval

Order	Family	Genus & Species
Carnivora	*Felidae*	*Felis serval*

Identification - One of East Africa's most beautiful cats, with very long legs, small head and very large, rounded ears. The coat colour is yellow/buff boldly and irregularly marked with black spots and bars. The chest and underparts are off-white. The tail is short and has alternate black and yellow/buff rings. The backs of the ears are black with a pronounced white spot towards the centre. In some regions servals with completely black coats are quite common, particularly in the Aberdares of central Kenya and the highlands of Ethiopia.

Distribution & Habits - Found throughout much of the region on savannahs and open grasslands as well as in light woodlands, around marshes and along forest edges. They hunt rodents, lizards, snakes and birds in areas of long grass, pin-pointing the whereabouts of prey by sound onto which they pounce with a final leap. Birds flushed from the grass are often caught in mid-flight with lightening reflexes and a leap into the air. They will also occasionally raid poultry stocks in and around human settlements. They are active in the early morning, late afternoon, evening and throughout the night. They live a solitary existence, maintaining territories up to 5 square kilometres in extent and, when not hunting, will rest up in subterranean burrows, natural crevices between boulders and in dense patches of vegetation. The senses of hearing and sight are very acute, while the sense of smell is less well developed.

KEY FACTS

Size	Length including tail: 110 cms (average).
	Height at shoulder: 60 cms.
	Weight: 13 kgs (average).
Breeding	Gestation: About 2.5 months.
	Young: Usually 2 or 3 (1-4).
	Sexual Maturity: 2 years.
	Births: No specific season.
Lifespan	About 12 years.
Lifestyle	Family: Solitary.
	Diet: Carnivorous.
	Main Predators: All large predators.
	Habitat: Open savannahs, marshes and woodland and forest edges.

Conservation & Status - Their numbers would appear to be stable over most of the region, but they are coming under increasing threat from the settlement of new areas by the expanding human population. They are also occasionally poached for their skins.

Greater Galago

Order	Family	Genus & Species
Primates	*Lorisidae*	*Galago crassicaudatus*

Identification - Also known as the Greater Bushbaby, this nocturnal primate has a rounded head, large ears, stout limbs and huge forward facing eyes. The coat is thick and fluffy and shows a wide variation in colour from overall grey with a white tip to the tail, to dark brown with a blackish tail tip. The darker animals tend to be found at higher altitudes and in damper habitats, lighter animals usually being found in drier, lower lying regions. The underparts are off-white and the short, pointed face has dark patches along the sides of the muzzle and around the eyes. The thick tail is longer than the combined length of the head and body, and is very bushy. They are expert climbers and have well adapted digits for this purpose.

Distribution & Habits - Found throughout Tanzania, southern Kenya and coastal regions of Somalia, in a variety of habitats including wooded savannahs, bush and scrub and mountain forests to an altitude of 3500 metres. They spend the daytime sleeping in hollow trees or dense foliage, becoming active at dusk and throughout the night. They feed on fruits, berries, seeds, insects, tree sap and small birds. Females build nests in hollow trees or dense vegetation in which to give birth. The young remain in the nest for the first 2 weeks of life, after which the females will carry them during their nightly foraging, provided the young have acquired the necessary growth and strength to hang on. They usually ride 'jockey style' on the mothers back. The senses of sight, hearing and smell are extremely good. Small family groups will maintain a territory several hectares in extent, making their presence known by depositing scent from a breast gland and by urinating on branches and tree trunks.

KEY FACTS

Size	Length including tail: 80 cms (average). The tail being approximately 60% of the total length. Weight: 1.5 kgs.
Breeding	Gestation: About 4.5 months. Young: 1-3 Sexual Maturity: 12 months. Births: All year round but with a peak prior to the onset of the rainy season.
Lifespan	Up to 14 years.
Lifestyle	Family: Singly or small family parties. Diet: Omnivorous. Main Predators: Large birds of prey and most large predators when on the ground. Habitat: Mountain forests, wooded savannahs, bush and thickets.

Conservation & Status - The biggest threat to the survival of the species is the destruction of suitable habitat. Numbers seem stable but are subject to occasional sudden reductions as a result of disease.

Lesser Galago

Order	Family	Genus & Species
Primates	*Lorisidae*	*Galago senegalensis*

Identification - Also known as the Lesser Bushbaby, this charming little primate is about half the size of the Greater Galago. They have a rounded head, short muzzle, large eyes and substantial pointed ears. The coat is soft and fluffy, grey over much of the body and tail, with a wash of yellow on the flanks and limbs. The underparts are whitish. They have dark eye patches and the insides of the ears are pink. They have a conspicuous light stripe extending from the forehead to the nose. The tail is longer than the combined length of the head and body and is bushy towards the tip.

Distribution & Habits - Distributed over much of the region in dry woodlands, bush and scrub, wooded savannahs and mountain forests to an altitude of 2000 metres. They are active throughout the night, spending the daylight hours sleeping in hollow trees, in forked branches or in self-built nest of leaves and twigs. Family parties will often sleep together in a single shelter. They feed on a wide variety of invertebrates, lizards, young birds and eggs as well as fruits, seeds and tree sap. The senses of sight, hearing and smell are very acute. Family territories are maintained by an old dominant male and, to a lesser extent by females, marking branches and trunks by the use of a scent gland on the breast and by the depositing of urine. The young remain in and around the birth nest for the first 10 days of life, often venturing short distances by riding on their mothers back. They grow quickly and attain full adult size by 4 months old.

KEY FACTS

Size	Length including tail: 43 cms (average).
	Weight: Up to 300 gms.
Breeding	Gestation: Around 4.5 months.
	Young: 1-2
	Sexual Maturity: About 9 months.
	Births: No specific season.
Lifespan	Up to 14 years.
Lifestyle	Family: Singly, pairs or family parties.
	Diet: Omnivorous.
	Main Predators: Large birds of prey, small cats.
	Habitat: Dry woodland, bush and scrub, wooded savannahs.

Conservation & Status - Their numbers appear to be stable, but they are under constant threat due to the spread of the human population and the consequent destruction of habitat.

Olive Baboon

Order	Family	Genus & Species
Primates	*Cercopithecidae*	*Papio anubis*

Identification - The commonest of East Africa's primates. A heavily built and powerful animal. The head is large with small, close-set, brown eyes beneath a protruding eyebrow ridge. The coat colour is a mixture of olive grey and black, becoming darker on the lower legs and feet. The underparts are off-white/grey. The long, pointed muzzle is black and devoid of fur. The ears are rounded and sometimes hardly discernable among the long fur of the head, particularly in the males, who develop a mane around the head and shoulders. The tail is of medium length, extending upwards for a few cms., before curving downwards at a sharp angle, giving a 'broken' appearance. They have extremely formidable canine teeth.

Identification & Habits - They are distributed throughout northern Tanzania, central and western Kenya and Uganda in a variety of habitats, including woodlands, savannahs and rocky areas. They live together in 'troops' varying in number from just a handful to 150 individuals, although around 35 would appear to be the average troop size. A distinct hierarchy exists among both males and females within each troop, maintained by alliances between members, often reaffirmed by mutual grooming. Females outnumber males within a troop by as much as 3 to 1. They feed throughout the day both on the ground and in trees, taking fruits, leaves, buds, grasses, insects, lizards and occasionally new born antelopes. Baboons can do untold damage to cultivated crops and, in some regions, are considered to be vermin. At night they roost together in trees or on rock ledges to avoid predation by Leopards and, to a lesser extent, other predators. A female will carry a new born beneath her during daily foraging, but as the youngster grows it soon adopts a jockey style position on it's mothers back.

KEY FACTS

Size	Length including tail: 130 cms (average)
	Weight: Up to 35 kgs. Females are around half the size of males.
Breeding	Gestation: About 6 months.
	Young: 1 only, twins very rare.
	Sexual Maturity: Females 4 years, Males 6 years.
	Births: No specific season.
Lifespan	Around 30 years.
Lifestyle	Family: Large family groups ('Troops')
	Diet: Omnivorous.
	Main Predators: Leopards & Lions.
	Habitat: Woodlands, savannahs and rocky areas.

Conservation & Status - In recent years their numbers may well have increased due to a reduction in the number of major predators such as the leopard. They are, however, continually in conflict with farmers and many are killed annually.

Yellow Baboon

Order	Family	Genus & Species
Primates	*Cercopithecidae*	*Papio cynocephalus*

Identification - Differs from the Olive Baboon in having a slender build and thin legs. The coat is short, yellowish/grey above with white/cream underparts. The long, pointed muzzle and the hands and feet are greyish. The head is round and the ears are small. They lack the mane around the head and shoulders which is such a distinctive feature of the male Olive Baboon. The small, amber coloured eyes are close-set. The eyelids are white/pink and play an important role in communication, particularly in displaying aggression. The tail is long and thin, extending upward for a short distance before 'breaking' downwards at a sharp angle.

Distribution & Habits - An animal of savannahs and woodlands in Tanzania and eastern and northern Kenya, where they forage both on the ground and in the trees. They feed on a wide variety of foods including, roots, tubers, seeds, leaves, grasses, fruits and many species of invertebrate. They live in large groups ('troops') each averaging around 30 individuals, dominated by a large male. Relationships are cultivated within the troop by all members, resulting in a well defined hierarchy. The baboons will often engage in lengthy bouts of mutual grooming which helps to cement these relationships. They are territorial animals with each troop collectively maintaining a range of up to 30 square kilometres, which they defend forcefully against neighbouring troops. Only the highest ranking males will mate with a female during the peak of her cycle, other high ranking males may mate with females at other times.

KEY FACTS

Size	Length including tail: 130 cms (average).
	Weight: Around 30 kgs. Females are around half the size of males.
Breeding	Gestation: About 6 months.
	Young: 1 only, twins very rare.
	Sexual Maturity: Females around 4 years, Male around 6 years.
	Births: No specific season.
Lifespan	Around 30 years.
Lifestyle	Family: Large family groups.
	Diet:Omnivorous.
	Main Predators: Leopards and Lions.
	Habitat: Savannahs and woodlands.

Conservation & Status - Numbers appear to be quite stable.

Patas Monkey

Order	Family	Genus & Species
Primates	*Cercopithecidae*	*Erythrocebus patas*

Identification - A long-limbed, agile primate, spending the vast majority of the daytime foraging on the ground in open areas. The coat is shaggy, the back and crown are rufous red whilst the underparts and limbs are white, often with a flush of yellow. The facial colour varies from grey/pink to black, with a distinctive black band extending along the line of the eyebrows. The males have brighter, more pronounced markings than the females or young and, in addition, often have a trace of grey on the shoulders. The long, thin tail is about the same length as the head and body, being rufous in colour and is carried high above the back when the animal is on the move.

Distribution & Habits - Found in northern Uganda, west and central Kenya and in isolated pockets of northern Tanzania. They inhabit dry, semi-desert regions, open savannahs, scrub and woodlands. They usually live in troops, often with a single dominant male and a number of females and dependant young, numbering around 25 individuals. Small bachelor groups may also be encountered. They forage and feed in a loose group on seeds, pods, leaves, grasses, fruits, berries and occasionally invertebrates. They can cause considerable damage to cultivated crops, being partial to maize, millet, bananas etc. as a result they are often severely persecuted by farming communities. During the night they sleep singly or in pairs high in the treetops. They have very acute sight and hearing and are constantly on the alert for possible attack by lions or leopards. They will defend their territory against neighbouring troops of Patas, but will tolerate the presence of Baboons and Vervet Monkeys within their territory.

KEY FACTS

Size	Length including tail: Males 140 cms, females 115 cms.
	Weight: Males 20 kgs, Females 11 kgs.
Breeding	Gestation: Around 6 months.
	Young: 1 only.
	Sexual Maturity: 3 years.
	Births: Peak between Jan & May.
Lifespan	Up to 20 years.
Lifestyle	Family: Troops of 25 average.
	Diet: Omnivorous.
	Main Predators: Lions and Leopards.
	Habitat: Semi-desert, savannahs and woodlands.

Conservation & Status - Numbers appear stable, under some threat from the spread of the human population.

Sykes/Blue Monkey

Order	Family	Genus & Species
Primates	*Cercopithecidae*	*Cercopithecus mitis*

Identification - A large, robust monkey with a thick coat of blue/black fur. The back, face and hind legs are flecked with silver/grey hairs and occasionally washed with olive, whilst the crown and fore-limbs are black. A band of stiff grey hair extends across the line of the eyebrows. The chest and underparts vary from greyish to black. The small ears are black edged with grey. The close-set eyes are amber. The tail is longer than the combined length of the head and body, thickening towards the black tip. There is considerable regional variation in coat and facial coloration. The Blue Monkeys found in the forests of Mount Kenya having a collar of white fur and a dark reddish back.

Distribution & Habits - A monkey of dense, moist forests of coastal and mountain regions, often to an altitude of 3500 metres. They are usually encountered in troops of around 10 individuals, feeding among the trees and occasionally on the forest floor. Their main food intake consists of leaves, flowers, berries, fruits, some insects and birds eggs and young. The troops usually consist of several related females and dependant young, with a dominant male. Female off-spring will usually remain within the troop on reaching maturity, males, however, are forced to leave and join other troops or establish their own harem. They are active throughout the day, remaining in the shade provided by foliage, and will avoid direct sunlight whenever possible.

KEY FACTS

Size	Length including tail: 130 cms (average).
	Weight: 6kgs (average).
Breeding	Gestation: 4 months.
	Young: 1 only.
	Sexual Maturity: Around 5 years.
	Births: No specific season, although there is some evidence to suggest a peak during the rainy season.
Lifespan	Unknown.
Lifestyle	Family: Small groups.
	Diet: Omnivorous.
	Main Predators: Leopard, Python and Eagles.
	Habitat: Dense forests.

Conservation & Status - Numbers appear to be stable at the moment, but they are under constant threat in some regions due to deforestation.

Black-Faced Vervet Monkey

Order	Family	Genus & Species
Primates	*Cercopithecidae*	*Cercopithecus aethiops*

Identification - One of the commonest of East Africa's monkeys. A very agile and slender primate with long limbs. The coat colour is subject to some regional variation, but is usually grey/brown with a wash of olive on the back, a naked jet black face surrounded by white hair on the cheeks and across the eyebrow line. The chest and underparts are whitish, the lower part of the limbs are greyish with black feet. The eyes are deep, rich brown, the tail, which is about the same length as the head and body, is greyish with a black tip. The small ears are black. Both sexes have brightly coloured genitalia, the male having a powder blue scrotum and a bright red penis. The males are about 20% larger than the females.

Distribution & Habits - A widespread species found in woodlands, forests, savannahs and areas of bush and scrub. They are active throughout the day and are usually encountered in troops varying in number from 6-50+, foraging in the trees and on the ground. The composition of the troops usually shows an even number of males and females, with up to 50% being juveniles. The size of territory maintained by each troop varies greatly, as a consequence of food availability. Their main food intake consists of fruits, berries, seedpods, small invertebrates, lizards and birds eggs and young. They will readily eat cultivated crops and in areas of intense farming have been almost exterminated by the farming community as a result. Vervet monkeys have a well developed system of both visual and vocal communication.

KEY FACTS

Size	Length including tail: 120 cms (average)
	Weight: 5 kgs (average).
Breeding	Gestation: 5-6 months.
	Young: 1 only.
	Sexual Maturity: 2.5 years.
	Births: No specific season.
Lifespan	Around 20 years.
Lifestyle	Family: Troops of 25 (average).
	Diet: Omnivorous.
	Main Predators: Leopards, large Eagles, Crocodiles, Pythons
	Habitat: Woodlands, forests, savannahs and bush.

Conservation & Status - Very widespread and numerous.

Black & White Colobus

Order	Family	Genus & Species
Primates	*Cercopithecidae*	*Colobus guereza (Eastern)*
		Colobus angolensis (Western)

Identification - This large black and white monkey is represented in East Africa by two distinct species, the Eastern and the Western. In both species the overall body colour is black, the differences between the two shows in their distribution and in the extent of white body fur. The Eastern Colobus has a black face surrounded by short white fur, has long, flowing white fur extending from the shoulders along the length of the body to the rump and a long black tail with a bushy white tip. The Western Colobus has a black face surrounded by long white fur, with flowing white fur on the shoulders only and a long, thin all white tail. It is worth noting that the extent of white fur on the body in both species is subject to considerable regional variation. Newborn animals are all white with pink faces.

Distribution & Habits - The Eastern Colobus if found in northern Tanzania, central and western Kenya and Uganda. The Western Colobus is found mainly in coastal regions of Kenya and Tanzania and in south-western Uganda. They inhabit dense mountain and coastal forests, acacia woodlands and wooded savannahs. They feed mainly on fresh leaves, fruits, bark, seedpods and some insects. They are active throughout the day but will usually rest up during the hottest, midday period. They spend very little time on the ground, usually making their way through the forests by leaping from tree to tree. They live in troops of 3-12 consisting of a dominant male with several females and offspring.

KEY FACTS

Size	Length including tail: 145 cms (average). The tail accounts for around half of the total length.
	Weight: 15kgs (average)
Breeding	Gestation: Around 6 months.
	Young: 1 only.
	Sexual Maturity: Males at 4 years, Females at 2.
	Births: No specific season.
Lifespan	Up to 20 years.
Lifestyle	Family: Troops of 3-12.
	Diet: Omnivorous.
	Habitat: A wide variety of forests and woodlands.

Conservation & Status - There has been a decline in numbers during recent years as a result of deforestation.

Lowland Gorilla

Order	Family	Genus & Species
Primates	*Pongidae*	*Gorilla gorilla graueri*

Identification - A large animal with immense strength. The head is large and the body broad, the arms are considerably longer than the legs. The face is flat with deep set eyes and large nostrils. All areas of naked skin on the face, hands, feet and upper chest are black. The mainly black coat is short and dense, compared with the longer, shaggy coat of the Mountain Gorilla. The males acquire a 'saddle' of silver hair across the lower portion of the back on reaching maturity, hence the name 'silverback' used when referring to dominant males. The males generally become greyer with age. On average females are only half the size of the males.

Distribution & Habits - An inhabitant of the lowland tropical rain forests of eastern Zaire. Although they will climb trees the vast majority of their time is spent on the ground, feeding on a wide variety of vegetation including leaves, fruits, ferns and tubers. They forage over home ranges of between 10-30 square kilometres in extent. They live in small family groups, usually consisting of a dominant male and occasionally subordinate males with several females and their offspring. Groups generally keep well clear of each other, but when they do meet the dominant males will display by calling and chest beating. During the early years of African exploration there were many tales related of the ferocity and aggression of gorillas, in more recent times studies have shown gorillas to be rather shy, placid animals unless unduly threatened. At night each animal builds a sleeping nest either on the ground or in a tree, a new nest is constructed every night.

KEY FACTS

Size	Height: 1.75 metres (average)
	Weight: Males 190 kgs (average),
	Females 95 kgs (average).
Breeding	Gestation: 8.5 months.
	Young: Usually 1, twins very rare.
	Sexual Maturity: Males around 10 years,
	Females around 9 years.
	Births: No specific season.
Lifespan	Around 30 years.
Lifestyle	Family: Small family groups.
	Diet: Omnivorous.
	Main Predators: Man and Leopards.
	Habitat: Tropical rain forests.

Conservation & Status - Endangered. Under constant threat from the destruction of forests by farming communities, to make way for cattle herds and from poaching.

Mountain Gorilla

Order	Family	Genus & Species
Primates	*Pongidae*	*Gorilla gorilla beringei*

Identification - An animal of powerful build and immense strength. The head is large, the body is broad and the arms are considerably longer than the legs. The face is flat, the forehead low, the eyes close together and deep set and the nose flat with large nostrils. The naked areas of skin on the face, hands, feet and upper chest are black. The black coat is longer and thicker than that of the Lowland Gorilla. The males acquire a 'saddle' of silver hair across the lower portion of the back on reaching full maturity, hence the name 'silverback' used when referring to dominant males. Males generally become greyer with age. Male are around twice the size of females.

Distribution & Habits - Found in the high mountainous terrain of south western Uganda and in neighbouring Zaire and Rwanda. They live in relatively stable family groups, each group being lead by a dominant adult male. Each group usually consists of the lead male with several females and their offspring. Any other adult males within a group will be subordinate and most probably be sons of the dominant silverback. These young males will eventually be forced to leave the group and establish harems of their own. Each group occupies a home range of 10-30 square kilometres in extent and will avoid contact with neighbouring groups whenever possible. If groups meet the dominant males will display by calling and chest beating, avoiding physical contact if at all possible. They feed on a wide variety of vegetation including leaves, buds, fruits, ferns and tubers. They are active throughout the day and at night each animal will construct a sleeping nest on the ground or in a tree. A fresh nest is constructed each night.

KEY FACTS

Size	Height: 1.75 metres (average)
	Weight: Males 190 kgs (average),
	Females 95 kgs (average).
Breeding	Gestation: 8.5 months.
	Young: Usually 1, twins very rare.
	Sexual Maturity: Males around 10 years,
	Females around 9 years.
	Births: No specific season.
Lifespan	Around 30 years.
Lifestyle	Family: Small family groups.
	Diet: Omnivorous.
	Main Predators: Man and Leopards.
	Habitat: High mountain forests.

Conservation & Status - Endangered. Under constant threat from the destruction of habitat and from poaching.

Chimpanzee

Order	Family	Genus & Species
Primates	*Pongidae*	*Pan troglodytes*

Identification - A robust and powerful primate, the chimpanzee is our closest living relative. Individuals vary greatly in size and colour but are generally 1 - 1.7 metres in height, weigh between 40 and 50 kilos and have a shaggy black coat. They have a sturdy body with well developed shoulders and arms, the legs are rather short. The face is flat with small deep set eyes, a small nose, a broad, deep upper jaw and a wide, narrow-lipped mouth. The ears are rounded and prominent. The facial coloration varies from pink in juveniles to black in full adults, with only a sparse covering of hair on the cheeks and forehead. They walk on the hind legs and the knuckles of the forelimbs.

Distribution & Habits - Restricted in distribution to the rain forests and woodlands of western Tanzania and Uganda. They generally live in small, loosely knit groups of 2 - 50 individuals. They are equally at home in the trees or on the ground, where they forage throughout the daytime, feeding on leaves, fruits, nuts, ants, termites and, on occasions, young monkeys, antelopes and birds. They often fashion crude tools from twigs in order to extract larvae and other food items from holes and crevices. During the night they sleep in nests built in the trees from branches and leaves, although on moonlit nights they will sometimes move about and feed. They are the most vocal of all primates and have an extensive vocabulary of sounds. Each group has a territory or home range of between 8 and 20 square kilometres. Encounters between neighbouring groups elicits much display and vocalisation prior to physical contact which, quite often, results in the death of rival members.

KEY FACTS

Size	Variable, 100 cms - 170 cms.
	Weight: 45 kgs (average)
Breeding	Gestation: 8 months.
	Young: Usually 1, twins occasionally.
	Sexual Maturity: Males 10 years, Females 9 years.
	Births: No specific season.
Lifespan	Around 40 years.
Lifestyle	Family: Small to large family groups.
	Diet: Omnivorous.
	Main Predators: Man and Leopard.
	Habitat: Forests and woodlands.

Conservation & Status - Very few chimpanzees remain in East Africa. Their continued existence depends on large tracts of forest being left undisturbed, but such areas are under constant threat of clearance by loggers and agriculturalists.